P9-EDK-094

Beyond the Green Revolution

The Ecology and Politics of Global Agricultural Development

Beyond the Green Revolution

The Ecology and Politics of Global Agricultural Development

Kenneth A. Dahlberg
Western Michigan University
Kalamazoo, Michigan

Plenum Press · New York and London

Library of Congress Cataloging in Publication Data

Dahlberg, Kenneth A
Beyond the green revolution.

Bibliography: p.
Includes index.
1. Agriculture and state. 2. Green revolution. 3. Agricultural ecology. I. Title.
HD1415.D273 338.1'8 78-11271
ISBN 0-306-40120-7

A Division of Plenum Publishing Corporation
227 West 17th Street, New York, N.Y. 10011

Printed in the United States of America

Preface

This book, which is the result of an intellectual odyssey, began as an attempt to explore and map the environmental and cross-cultural dimensions of the continuing spread of the green revolution—that package of high-yielding varieties of grain, fertilizers, irrigation, and pesticides that constitutes the core of modern industrial agriculture. In the process of traversing the terrain of several intellectual traditions and cutting through various disciplinary forests and thickets, a number of striking observations were made—all leading to two sobering conclusions. First, most intellectual maps dealing with agriculture fail to recognize it as the basic interface between human societies and their environment. Because of this, they are little better than the "flat earth" maps of earlier centuries in helping to understand global realities. Second, when agriculture is analyzed from a global perspective that takes evolution seriously, one sees that the ecological risks as well as the energy and social costs of modern industrial agriculture make it largely inappropriate for developing countries. Beyond that, one can see a great need within industrialized countries to develop less costly, less risky, and more sustainable agricultural alternatives.

Early in the journey it became clear that conventional disciplinary approaches were inadequate to comprehend the scope and diversity of global agriculture and that a new multilevel approach was needed. It also became clear that any new approach would have to try to correct certain Western biases and blind spots. Despite claims of neutrality and universality, modern science and technology are outgrowths of Western civilization, whereas modern agriculture is a product of the temperate zones. Both inevitably involve a number of culturally and environmentally specific elements that are neither neutral nor universal. While some of these have been exposed through unsuccessful attempts to "transfer" various technologies (a concept which is itself based on the idea of neutrality and universality), the real challenge is to sort them out analytically.

A new approach—called contextual analysis—that attempts to deal with these various difficulties is outlined here. In seeking to take evolutionary and historical processes seriously and to capture both their breadth and depth, three different levels of analysis are used simultaneously. These are used to sort out different natural, social, and technological processes and events according to three different time-frames. The underlying conviction is that only by seeking to understand real people, groups, plants, technologies, etc., in their specific and evolving context can we overcome the weaknesses of conventional approaches. This new approach naturally seeks to be holistic but in the broadest sense. At the first level of analysis, diverse disciplinary materials are brought together; at the second level, bridges are sought between the social sciences, the natural sciences, and the humanities; and at the third level, a synthesis of natural and human evolution is sketched.

The latter part of the journey has involved a going back and forth between the intellectual tools developed at each level and their additional shaping and sharpening on the anvils of history and the grindstones of present experience. The goal has been not only to understand the past and the present better, but to outline the future and its tasks and priorities more clearly. The overarching priority is to keep our evolutionary options open. Doing this will not be easy, particularly since many of our current paths lead to evolutionary dead ends. New and alternative paths have been sought and sketched out. Over the next century, basically different strategies will be needed for the industrial as compared to the developing countries. This implies a challenge to current definitions of progress and suggests the need for creative new visions. Such strategies must be based on comprehensive concepts of societal development that are rooted in specific environments. They cannot be derived from sectoral extrapolations. The future of agriculture must be understood in terms of larger evolutionary trends and limits, infrastructural and institutional developments, and cultural changes. The details of these broad concepts and alternatives and the ways in which they challenge current agricultural research and development strategies are best left to the book itself.

Since the book is meant for several audiences, some guidance as to which chapters are most relevant for each reader may be in order. Chapter 1 is a theoretical introduction to both the time-frames employed and to better ways to sort out the Western dimensions of science and technology. It is of primary interest to academic colleagues in the social and natural sciences. Chapter 2 gives an overview of agricultural history, describing how we got where we are, while at the same time trying to identify the major shifting parameters of recent centuries. It should be of general interest but could be skipped by those with a particular interest in contemporary and future problems. Chapter 3 contains the central critique of the green revolution. Chapter 4 outlines the influence and momentum of those general institutions, infrastructures, and policies that have grown up during the past decades and that will have to be dealt with if alternatives are to be sought in any serious manner. It also contains a fundamental critique of conventional theories of economic development. Chapter 5 deals with new approaches to the future, essentially trying to sort out various evolutionary, developmental, and policy priorities. It should be of use to policy makers in both industrial and developing countries, whatever their sector (domestic or international), and should make a contribution to the emerging field of risk assessment. Chapter 6 applies the insights gained throughout the book to the problems of agricultural development, suggesting the ways in which alternative approaches can be developed. It should be of particular use to development planners and theorists, especially those interested in appropriate technologies and rural development. Chapter 7 reflects on the journey, the magnitude of societal changes and global transformations that are visible on the horizon, and the kinds of political skill and intellectual creativity that will be required to cope with them.

The writing of this book started during a sabbatical year at what must be the perfect combination of physical and intellectual environments: the Institute for the Study of International Organization (ISIO) at the University of Sussex, England. Robert Rhodes James, who was then Director of ISIO, provided both gracious hospitality and practical guidance. Brian Johnson, who later became Director of ISIO, not only read drafts of the early chapters but has also provided continuing intellectual stimulation. A special intellectual debt is owed to Egbert de Vries. More than any other individual, he has helped me understand the complexity of interactions between natural and social systems at all levels. In a free-floating kind of seminar, I have been privileged to draw upon his extensive experience, keen observation, and gentle humor. Another who has been of special assistance is Andrew Pearse. As director of the project on "the social and economic implications of large-scale introduction of new varieties of foodgrain" of the United Nations Research Institute for Social Development, he supplied me with detailed country, regional, and topical studies.

A number of colleagues and friends helped by reading and commenting on various chapters. These include Stillman Bradfield, Joseph Collins, Charles F. Heller, Janos Horvath, Charles O. Houston, Brian Johnson, Maynard Kaufman, Dale H. Porter, William A. Ritchie, Zdenek J. Slouka, and Egbert de Vries. Their suggestions have improved the organization of the book and have helped me avoid various errors or questionable interpretations. Those that remain are obviously my responsibility. The vital supporting task of typing the various drafts of the book has been carried out expertly and cheerfully by Dorothea Bradford. Finally, whatever stylistic felicity the book has is due to the careful (and sometimes painful) editorial paring of my wife Helen. That, plus the various other important ways she has helped to sustain me, lead me to dedicate the book to her.

KENNETH A. DAHLBERG

Acknowledgments

For permission to use copyrighted material in this book, grateful acknowledgment is made to the following sources:

For Figure 2: From *Perspectives in Ecological Theory* by Ramón Margalef. Copyright © 1968 by The University of Chicago.

For Figure 4: From *Too Many* by Georg Borgstrom. Copyright © 1969 by Macmillan Publishing Co., Inc. Copyright© 1969, 1971 by Georg Borgstrom.

For Table 6: From *Managers of Modernization* by Leslie L. Roos, Jr. and Noralon P. Roos. Copyright© 1971 by the President and Fellows of Harvard College.

For Table 7: From *Before Nature Dies* by Jean Dorst. Copyright © 1965 by Delachaux & Niestlé S. A. Copyright© 1970 by William Collins Sons & Co. Ltd. Reprinted in the United States by permission of Houghton Mifflin Company.

Contents

1

On the Ecology of Theories

Time is nature's way of making sure that everything does not happen at once.

Anonymous

In a striking passage, the ecologist Ramón Margalef points out how both the perceptions and the theories of different ecologists are strongly shaped by the particular kind of environment they study:

> Ecosystems reflect the physical environment in which they have developed, and ecologists reflect the properties of the ecosystems in which they have grown and matured. All schools of ecology are strongly influenced by genius loci that goes back to the local landscape. "Desert" ecologists, working in arid countries where weather fluctuations exert a controlling influence on poorly organized communities, would hardly accept as a suitable basis for ecological theory the points of view put forward in the preceding chapter. . . . The mosaic-like vegetation of Mediterranean and Alpine countries, subjected to millennia of human interference, has assisted at the birth of the plant sociology school. . . . Scandinavia, with a poor flora, has produced ecologists who count every shoot and sprout. It is a pity that the tropical rain forest, the most complete and the complex model of an ecosystem, is not a very suitable place for the breeding of ecologists. And it is only natural that the vast spaces and smooth transitions of North America and Russia have suggested a dynamic approach in ecology and the theory of climax. In such areas, the concept of succession, one of the great and fruitful ideas of classical ecology, was best formulated.[1]

[1]Ramón Margalef, *Perspectives in Ecological Theory* (Chicago: University of Chicago Press, 1968), pp. 26–27.

One of the root dilemmas of scientific thought is suggested here: how to reconcile such a variation in the theories within a field (which appear to accurately reflect and explain the environment being studied) with the regularity, uniformity, and universality generally postulated in conventional models of science? There are growing indications that these models and their highly abstract descriptions of what theories *should* do are not very accurate when it comes to describing how theories are actually generated and how they operate.[2] A close look at scientists' descriptions of the evolution of their major contributions reveals a picture quite different.[3] Indeed, some scientists have argued that scientific creativity is much more akin to artistic creativity than the standard models would suggest.[4] With this in mind, let us examine the "ecology of theories"—the ways in which theories evolve, flourish, adapt, or die in the larger environments around them.

The growth of what may be termed major theories[5] is generally as follows: the theorist, explorerlike, immerses himself in a lengthy examination of the specific fields he is interested in as well as of the surrounding areas. These explorations help make him sensitive to basic interactions and changing parameters; they also may raise questions that have been inadequately dealt with in prior theories. Variations between cultures and historical periods, as well as the personal perceptions and social values of individual theorists, make it hard to generalize about the struggle to comprehend and describe a chosen area; however, it seems safe to say that any major theory is a blend. It is in part a reflection of the environment and in part the theorist's interpretation of it. In the short run, the "success" (acceptance) of the theory seems to depend as much on the receptivity of the then-current

[2]Much of this has come out of the work of Thomas S. Kuhn, *The Structure of Scientific Revolutions* (Chicago: University of Chicago Press, 1962) and the debates and controversy surrounding it.

[3]See, for example, James D. Watson, *The Double Helix* (New York: Atheneum, 1968).

[4]Jacob Brownowski, *Science and Human Values* (New York: Harper & Row, 1972). See particularly Part I, "The Creative Mind."

[5]What is suggested here is a continuum ranging from world-view (a broadly based cultural phenomenon) to paradigm (a group of assumptions and values shared by a specific social group—scientists) to major theories (those that define basic boundaries or relationships) to derivative theories (those that flow from and are dependent upon major theories) to concepts, etc. These distinctions are made to indicate the general level of analysis we are concerned with in this chapter.

intellectual climate and the support (or opposition) of various academic groups as upon its intrinsic soundness.

To the degree that such a major theory is "successful," it often engenders and becomes entwined in a variety of vested interests. Professional associations are created, departments and courses are established, reputations are built, buildings and equipment are purchased, textbooks are written, and so on; by that point the theory is no longer purely and simply a theory but is inextricably woven into the lives, institutions, and interests of both the academic and the outside world. For a new theory to challenge what has become the orthodoxy of the old is a difficult, often generation-long task.[6] Not only are there the vested interests associated with the old theory, but its advocates and practitioners often have the ability to modify the environment under study to better fit "their" theory. Psychologists, for example, seem to have succeeded in breeding a laboratory rat well adapted to "proving" their stimulus-response theories.

Another support for existing theories lies in the profound influence of human habit. In a relatively stable environment, habits are quite useful; in a rapidly changing world, they can be disastrous.[7] The dangers of applying habitual modes of thought to the dilemmas posed by nuclear weapons and their proliferation are well documented.[8] The dangers of applying habitual thought to that mixture of problems called *the environmental crisis* are equally obvious. Even so, the difficulties that individuals face in changing their habits and modes of thought are small compared to those faced by institutions. And since today's academic research is heavily institutionalized and routinized, it is not surprising that there have been few large or imaginative attempts to reexamine the foundations of disciplinary science to see if its underlying structures and assumptions are still sufficient to deal with the real world.

[6]See Kuhn and also the section on "Generations and Events" in Benjamin Ward, *What's Wrong with Economics* (New York: Macmillan, 1972).

[7]Given the vital importance of habit to the healthy functioning of both individuals and societies, it is curious that the typical technocratic assumption is that people should adapt to rapid technical change, rather than vice versa. However, it must be recognized that even if priorities are reversed, the current momentum of rapid technological change is such that, short of catastrophe, it will slow down only gradually.

[8]Jerome D. Frank, *Sanity and Survival* (New York: Random House, 1967).

THE GROWTH OF CURRENT THEORIES

Since theories must be considered in terms of the actual situations from which they grew and on which they act, and since modern science is primarily a creation of Western society, to what degree must it be said that modern science is really Western science and that its theories are appropriate only for Western societies? Phrasing the question so starkly elicits the almost instinctive response that modern science is *universal*. Why? Because one of the deepest underlying *articles of faith* of modern science is in the universality—both spatial and temporal—of its laws and theories.[9] What is argued here is that current scientific theories are neither wholly universal nor wholly specific to Western society; they are mixtures that vary from field to field.[10] Figure 1 illustrates this. It suggests that there are increasingly large cultural biases (both Western and non-Western) as one moves from a rather abstract physical field like chemistry to a social area like economics and then on to a field like agriculture, where specific environments and cultures interface. Different geometric forms have been used to depict the different cultural and environmental influences that shape Western and non-Western theories and understandings. Figure 1 also

[9]For a discussion of this and other beliefs associated with modern science and technology, see Kenneth A. Dahlberg, "The Technological Ethic and the Spirit of International Relations," *International Studies Quarterly* 17(March 1973):55–88. Also see Robin Horton and Ruth Finnegan, eds., *Modes of Thought: Essays on Thinking in Western and Non-Western Societies* (London: Faber and Faber, 1973), for an interesting discussion of these questions as they relate to anthropological studies.

[10]There are a host of philosophical questions and problems associated with this assertion—which I do not propose to deal with here. Some comments may help with a few of these, however. Variations appear both according to the subject matter and according to the level of abstraction. Physics generally claims the most universality. However, even there, one can ask about the status or existence of the law of gravity before the "big bang" that created the universe—if one accepts the latter theory. What this suggests is that different fields claim universality only "within limits"—whether those limits are defined in terms of the types of phenomena (physical, organic, social, etc.), in terms of normal operating conditions (particularly within certain temperature limits), or in terms of the temporal period covered (the example above). Also, as the shifts from Newtonian to Einsteinian conceptions illustrate, the claim is that succeeding theories represent closer *approximations* to the phenomena—which are assumed to be universal; even here one can note variability in that it was because Einstein felt that length and mass were the cardinal variables in physics that he responded to the Lorentz contraction formula (which appeared to threaten the cardinality of length) by "saving" this cardinality through a conception that postulated the "contraction" of time—see Nicholas Georgescu-Roegen, *The Entropy Law and the Economic Process* (Cambridge, Mass.: Harvard University Press, 1971), p. 104.

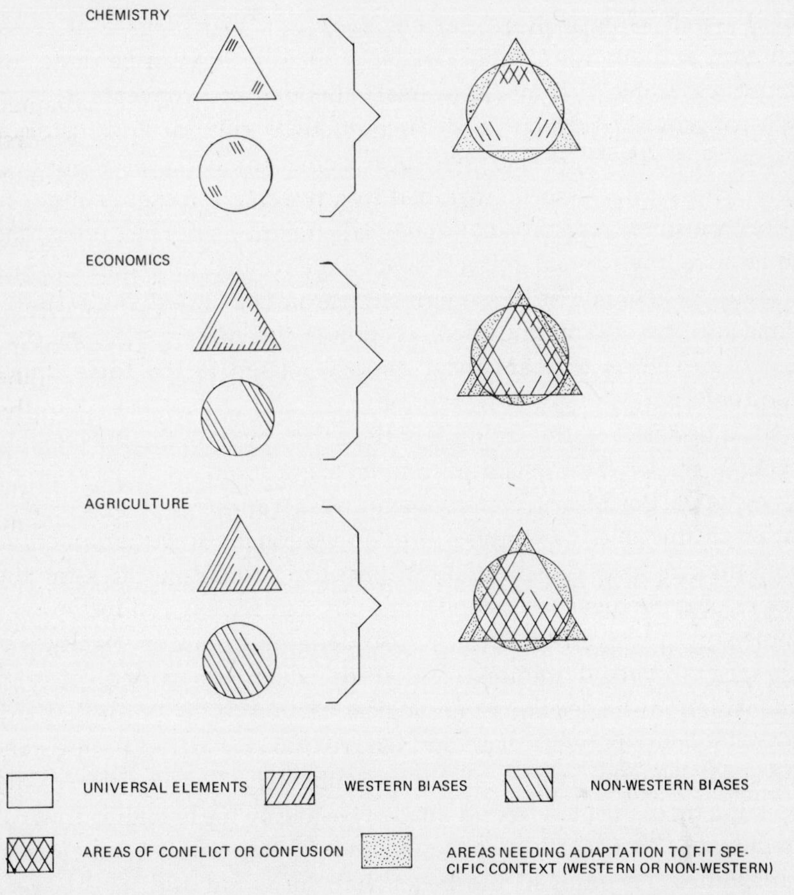

Figure 1. Degree of universality and compatibility of various modern fields.

suggests something regarding theories that applies even more to Western institutions and technologies: they are genuinely applicable to non-Western situations only if they are significantly changed and adapted.[11]

These points, if taken seriously, indicate that different research strategies (not to mention policy approaches) are necessary in any

[11]This follows the discussion in Brownowski, pp. 18–19, in which theory and practice are seen as being on a continuum. Thus, there is less cultural admixture the more one moves toward theory, and more as one moves toward the practical. In the following discussion, what is said to apply to theory as regards cultural admixtures or distortions applies even more to the technologies and institutions that attempt to apply these theories.

field—such as agriculture—where there is a high degree of cross-cultural and environmental variability. Before this can be done, we must reexamine current "universal" theories and concepts to more self-consciously sort out and discount their cultural and historical biases. Since there are also physical and environmental variations over time, these must also be included in any serious reexamination. In short, the growth of current "universal" theories must be understood in terms of their actual cultural-space-time contexts. This means first tracing the general rise of modern science and its underlying assumptions and then examining the evolution of the specific theories, concepts, and ideas that are most directly related to the topic under consideration—the green revolution.

To understand the strength of the assumption of the universality of modern science (an assumption open to question), we must examine its historical buttresses. One prime buttress was the gradual transmutation of the idea of progress from an essentially Christian doctrine with the emphasis on spiritual concerns to a secular one stressing the hopes of rationalists for establishing through reason and technology some sort of heaven on earth.[12] Also deriving from Christianity was the idea of man's dominion over nature, which was carried over and intensified in the secular technological version to the point that any ethical relation between man and nature was considered nonexistent (at least until the rise of all those "subversive" ecologists).[13] The opening up of the New World, made possible by the breaking down of traditional "flat earth" concepts and by growing technological capabilities, encouraged the belief that the world and nature were man's to conquer.

The belief that Western society was superior in all ways (intellectual, cultural, religious, technical, and so forth) showed itself in the way territories and peoples were conquered and converted willy-nilly to the uses and beliefs of their subjugators. These centuries-old beliefs

[12]See Ernest L. Tuveson, *Millenium and Utopia: A Study of the Background of the Idea of Progress* (Berkeley: University of California Press, 1949).

[13]See Lynn White, Jr., "The Historical Roots of Our Ecological Crisis," in *The Subversive Science*, edited by Paul Shepard and Daniel McKinley (Boston: Houghton Mifflin, 1969). At least with the Christian view, nature was a part of God's creation and thus was due some respect. In fact, within the Christian tradition it was possible for "radicals" like Saint Francis of Assisi to suggest the equality of all living creatures.

have been a bit shaken by decolonialization and somewhat muted in the courtesies required between "sovereignly equal" countries, but they are still visible in the tendency of the Western countries (including the USSR) to assume that their values and development models are the most appropriate ones for the non-Western world.[14] Things like foreign aid, technical assistance, and the Peace Corps all contain the basic assumption that Westerners have a great deal to tell the developing countries; the idea that there might be a return flow of aid and useful information—a Peace Corps in reverse—would be met by most people with a polite laugh. Why? Obviously, the knowledge that, say, an Indonesian rice grower might have to offer is so colored by his particular culture and environment that it would be extremely difficult for him to isolate those points that might be of use to an impoverished rice grower in Mississippi. His natural tendency, given enough power and influence, would be to suggest that the Mississippian restructure his rice paddies and his cultivation techniques to fit more closely those in Indonesia. If this is logical and natural, then why is it so difficult to see that the reverse situation (the more likely one) is the same, with the Westerner trying to alter the practices and environment of the non-Westerner so that they will correspond more closely to his own Western experience and perception of the world? Surely the answer lies largely in the Westerner's belief in the superiority and universality of his science and technology.

The progressive specialization of Western societies is another broad historical trend that has had considerable influence on the structure and content of modern science. Undergirding this trend is a mode of thinking that assumes that

> by analyzing and specializing work along functional lines—whether that work is to be done by man or machine—the most efficient use will be made of available resources. In this mode of analysis, the "factors of production" are all defined as part of, but subordinate to the total production process, and it is for this reason that men (labor) and the environment (resources)

[14]Indeed, as one commentator has put it only half facetiously, "The assumption that development is a *generalisable* concept must be seen in this context. It is far more potent than the crude instruments of 'neo-colonialism.' It is the last and brilliant effort of the white northern world to maintain its cultural dominance in perpetuity, against history, by the pretence that there is no alternative." Address by John White, "What Is Development? And for Whom?" given at the U.K. Quaker Conference on "Motive Force in Development," April 1972, p. 4 (my italics).

> are molded to fit the production process rather than vice versa. Another basic aspect of this mode of analysis is the strong tendency toward standardization. Since the production process is broken down into a series of work units, or functions to be performed, it is but a short step to the idea that similar functions should be standardized. This of course means that the "functioners" (men and/or machines) must also be standardized, but this is a "small" price to pay for the greater efficiencies afforded through mass production and the interchangeability of parts.[15]

The application of functional specialization to Western societies has been a slow and varied process. In the economic realm, it moved from separate machines like the printing press (16th century) to individual enterprises (17th and 18th centuries) to the establishment of national economies (19th century) to what is now called the world economy. In politics, its application (through the varied forms of bureaucracy) has roughly paralleled the changes in the economic realm, undoubtedly because of their many mutual links. In other areas (religious affairs, military matters, the academic world), its application has been more varied in time, place, and extent.

The specialization that so dominates modern science has resulted from some fascinating interactions between scientific theory and technology. At times, scientific theory seemed to be spurred by technological developments (i.e., the telescope); at other times, the development of new theories led to the establishment of huge and complicated technological and bureaucratic systems (the theory of relativity leading to the Nuclear Establishment). While scientists have searched for progressively more abstract and general theories and laws, more and more specialization in thought and technology has been required to do this. It is rather like the average symphony goer who enjoys the works of Beethoven, Brahms, and Britten but is dependent for their enjoyment upon the specialized abilities of the hundred-odd members of the orchestra. Similarly, the work of the great scientific theorists of today rests upon a wide foundation of specialized intellectual and technical work and upon complex institutional linkages between industry, governments, and academia.[16]

As indicated earlier, the growth of such institutional complexes leads to a variety of vested interests that make the introduction of new

[15]Dahlberg, pp. 62–63.

[16]J. Robert Oppenheimer, *The Flying Trapeze: Three Crises for Physicists* (London: Oxford University Press, 1964), pp. 2–3.

scientific theories difficult. Additional difficulties include the specialization of scientific thought through the development of academic disciplines and their infusion with a particular mode of thinking: reductionism. There are a variety of reasons for the ascendancy of reductionism. It is in part a logical extension of the search for truth through science—looking for universal underlying causes for actions and behavior. It also rose out of the intellectual backlash to the last serious, but premature, attempt to construct a general theory inclusive of both natural and social sciences, the attempt made by the Social Darwinists.[17] At almost the same time, the disciplinary specialization championed by the German universities became the model for university reform throughout the world. Finally, academic specialization and reductionism were bolstered by the democratization of Western societies (progressing at quite different rates in different countries) and the increasing concern of universities with practical matters, ranging from the establishment of the "agricultural and mechanical" land-grant colleges in the United States to an increasing number of professional schools elsewhere.[18]

Before going on to some of the specific concepts and theories related to global agriculture and the green revolution, one final Western institution that has markedly shaped their evolution and content has to be discussed: the modern state system. Without going into its complex religious, political, and intellectual sources, it is clearly permeated with Western concepts of law, contract, and status. It has remained the dominant political institution in the Western world even though its base of legitimacy has shifted from religion and monarchy to nationalism and democracy. The succession from colonialism to decolonization has transplanted the state to the non-Western world, and its leaders there preach its virtues and seek its firmer establishment through what should be called *state nationalism.* This involves their attempt to make the state the primary object of loyalty by supplanting traditional loyalties to village, tribe, region, caste, religion, etc. This is done through mass mobilization, education, and telecommunications.

[17]See H. Stuart Hughes, *Consciousness and Society* (New York: Knopf, 1958).

[18]Attempts to counter these trends and to give some strong institutional frame to holistic or synthesizing theories not only have encountered institutional opposition but generally have failed to try to bridge the gap between the natural and the social sciences. Most attempts have been *within* the natural *or* the social sciences.

The easy availability of military technology means that "modernizing" elites can in most cases squash any serious threat to their regime, particularly when it comes from traditional sources.[19] They have strong vested interests in the state through both their training and their beliefs and (perhaps most of all) through the perquisites and powers of office. The strength of these vested interests can be measured by the meager results of attempted mergers of states (the Mali Federation and United Arab Republic, for example). While an interesting case can be made that any significant cultural renaissance among non-Western peoples might transform the nature of the state and the state system in those areas,[20] it seems reasonably clear that such changes would be short-lived without some significant changes within and among Western states themselves.

Pressures for such changes do appear to be building, especially in regard to the international economic order, the main complementary support system of the modern state. While the 18th and 19th-century doctrines and practices of laissez-faire, free trade, and imperialism have been modified into what is called development assistance or neocolonialism (depending on who is talking), the structure of the international economic system has remained clearly Western and, as the UNCTAD (UN Conference on Trade and Development) debates have brought out, benefits primarily the rich industrial countries. Any major movement toward a "new international economic order" will require not only political pressure from the non-Western world but a rethinking of economic theory. The latter must start with a reassess-

[19]Indeed, real threats appear to come more consistently from other "modernizing" sectors or groups.

[20]See Adda B. Bozeman, *The Future of Law in a Multicultural World* (Princeton, N.J.: Princeton University Press, 1971). Briefly, she argues that the possibility of such a transformation exists because there are deep and fundamental differences in the "general thoughtways and value systems" of the five different cultural zones that she identifies (Western, Islamic Middle East, Africa south of the Sahara, India and Indianized Asia, and China). These differences are only papered over by the "universal" language of international law and organization (which is clearly Western in origin). For example, the idea of contract—which has been writ large in international law as the law of treaties—is based on Western notions regarding individualism, good faith, and literacy. On the one hand, a major cultural renaissance in any of these areas would inevitably challenge these Western forms. On the other, effective cooperation at the international level will require a genuine appreciation of these different thought ways, values, and working vocabularies.

ment of the historical origins of various economic doctrines. The historical links between religion, laissez-faire capitalism, and growth of the middle class received attention in the works of Max Weber and R. H. Tawney.[21] Since then, a growing body of work has shown that the doctrines of free trade have been consistently championed by those countries in the strongest trading position: England in the 19th century and the United States in the 20th.[22] It is interesting that countries that have improved their relative trading positions—in particular the United States, Germany, and Japan—made strong use of the "infant industries" argument to protect what might be called their economic sovereignty during the early period of their growth.[23]

Such a reassessment will give further insight into the ways in which economic theory has influenced agriculture at both the national and the global levels. The broad outlines can be found in Adam Smith's *Wealth of Nations,* which is primarily concerned with trade and manufacturers. As his brilliantly laid-out model was progressively realized in an age of burgeoning industries, rapidly growing cities, and expanding middle classes, agriculture declined and became rather passé. Even today, both popular culture and conventional economic wisdom place agriculture in a Cinderella-like position—an indispensable but definitely lower-status stepsister. A number of things beyond serious droughts and famines suggest that agriculture should have a much higher intellectual and academic status. There is already important work showing that the Industrial Revolution was preceded and in large part made possible by both an agricultural revolution and a commercial revolution.[24] The fact that conventional economic theory has not been reworked to include these important historical points nor changed in response to the general failure of the industrial model to aid

[21]Max Weber, *The Protestant Ethnic and the Spirit of Capitalism* (New York: Scribners, 1958), and Richard H. Tawney, *Religion and the Rise of Capitalism* (London: Murray, 1926).

[22]See for example, Karl Polanyi, *The Great Transformation* (Boston: Beacon Press, 1957)

[23]It is also interesting that they did not hesitate to deal with weaker neighbors (Mexico, Austria, or Korea, respectively) in the same manner that they objected to on the part of stronger powers.

[24]On the agricultural revolution, see particularly Eric L. Jones, ed., *Agriculture and Economic Growth in England, 1650–1815* (New York: Barnes and Noble, 1967), which places the introduction of more intensive methods in the period 1660–1750 (which was also a period of population stability). On the importance of the commercial revolution, see Brian Johnson, *The Politics of Money* (New York: McGraw-Hill, 1970).

the development of the Third World can be seen as a concrete demonstration of the momentum of established theories and of the resistance of various vested interests to significant change.[25]

The magnitude of change needed to shift to an agricultural-ecological model of development is detailed later on; at this point it is sufficient to observe that as long as the standard academic approach to agriculture is maintained, there will be the general Western biases to overcome as well as the fragmentation of disciplinary specialization. Additionally, whatever synthesis of materials is attempted in the field risks being distorted by the dominant position of agricultural economists. Their belief in man's dominion over nature (leaving it as an almost residual category of analysis), their curious ambivalence toward peasant farmers (who are seen as irrational if they don't behave like Western "rational economic" men), and their systematic lack of concern with the environmental costs of agriculture (encouraged by an emphasis on universal mathematical measures) will all be changed only with reluctance—and even then only when a new approach is developed that clearly shows previous inadequacies, better integrates currently fragmented materials, and includes the real social and environmental costs and benefits of agriculture.

TOWARD A THEORY OF CONTEXTUAL ANALYSIS

The rest of this chapter is devoted to outlining the major dimensions of what I call *contextual analysis;* this outline is intended to be an introduction to a rather different approach (used throughout the rest of the book) for analyzing the green revolution. It is not in any way the formal presentation of a theory—that has to wait until the basic insights of this new approach are applied to several substantive issues to see how fruitful they are (or are not).

Contextual analysis, as a term, hopefully indicates the dual thrust of this approach. First and foremost, it means an attempt to get closer to the full richness and complexity of whatever real-world situation is being examined. In this sense it is holistic. Second, it is analytic, and concerned—as far as it is possible—with general trends and compari-

[25]Two recent critiques by economists detail the theoretical and sociological reasons for this. See Nicholas Georgescu-Roegen and Benjamin Ward.

sons. The major burden of the preceding section was that theories must be understood in terms of the environment, the values, and the particular historical contexts out of which they grew. For the analyst, the moral would seem to be that we had better be more self-conscious about the way our personal and cultural values influence our analysis, and that this self-consciousness has to be based on the kind of sensitivity to variation (environmental, personal, or cultural) needed to understand the complexities of any real-world situation. Contextual analysis stresses continual going back and forth between values, theories, and context so that their contours and interactions become somewhat clearer. Whether you see it as a dialectic or a feedback process, this kind of approach is needed to overcome one of the weaknesses of current "universal" theories, namely, that only a prolonged period of dissatisfaction with a theory's lack of applicability to particular situations will eventually raise questions about one or more of its basic parameters.

It is instructive in this regard that physicists have become more conscious of the costs and benefits of particular modes of analysis only since the Heisenberg experiments. Since then, they have had to frame their experiments with a careful eye to what they can or cannot observe simultaneously and therefore to what kinds of cause–effect relationships can or (more importantly) cannot be postulated.[26] The social sciences have been less concerned with estimating the costs and benefits of their methodologies and analytical modes. They have tended to use comfortable positivist assumptions that only measurable things constitute knowledge and that since the measures are "universal," the phenomena themselves, regardless of history or location, can be understood and compared through the manipulation of the measures,

[26]As Oppenheimer, p. 51, phrases it: "The important point is that it is not merely that we do not always know everything that in classical mechanics we thought we could know, like the position and momentum of an object; if that were so, you could say: 'Well, I know its momentum and I will suppose that it is distributed somehow over different possible positions and I will calculate what I'm interested in and take the average.' But you must not do that. If you suppose that an object whose momentum you have determined by experiment has a distribution in positions, no matter what the distribution, you will get the wrong answer. It is not that you do not know it; it is that it is not defined. The experiment which gives you the momentum forecloses the possibility of your determining the position. If you welch on it and say, 'Well, I want to know the position in the first place,' then you can, but then you lose the knowledge which the earlier experiment had given."

that is, through statistics.[27] Contextual analysis suggests that measures and methods of analysis must be related to both the specific context and the basic research questions asked. Costs and benefits take on great importance. The uncounted costs of conventional approaches—that is, the parameters, research questions, and interactions that have been ignored or neglected—are discussed briefly below, and detailed examples are brought out all through the book. Equally, there is a brief sketch of the basic parameters of contextual analysis; its costs and benefits will become apparent later when it is applied to agriculture and the green revolution.

Let me describe two ways of producing clothing to indicate the kind of theoretical and methodological differences found between universal and contextual approaches. The former is analogous to a clothing factory that can turn out only Size 42 suits; some variety can be introduced by using different disciplinary cloth and buttons, but the suits will still fit only a small percentage of the world's men, even though the worldwide average size may be 42. In addition, the Western social habits and values inherent in a suit are obvious. A nomadic Somali tribesman will find the suit irrelevant to both his living and his work requirements as well as his tastes, even if it fits perfectly. Contextual analysis functions like a tailor who begins with a general design but is very aware that fitting it to a specific individual and his needs is a necessary and critical part of his task. The actual differences may be slightly less than those in the analogy, but the basic difference in conception is clear.

Locations in space and time are two fundamental dimensions that must be included if you are to get close to the real-world context of anything. Neither has been consistently dealt with in social science literature. Space has generally been conceived of as jurisdictional space, that is, as the legal extent of various local, regional, or international bodies. There is often a rough correspondence between these

[27]One is impressed when comparing the literature on ecological methodologies with that of international relations methodologies that ecologists self-consciously "tailor" their methodologies to their particular experiments (and this usually after a careful cost/benefit analysis). A typical abuse of international relations methodologists is to take standard statistical formulas and apply them to whatever statistical data may exist [see particularly the work by Bruce R. Russett, *International Regions and the International System* (Chicago: Rand McNally, 1967)] rather than tailoring their materials and methods to carefully worked out research and/or theoretical questions.

jurisdictional hierarchies and physical size, but there can also be great variation within a given jurisdictional category; Luxembourg and Brazil, for example, both fall within the state category.[28] Geopolitical theorists have tried to deal more precisely with the purely physical parameters of size, distance, and location as they relate to politics (and most particularly to power); however, as the Sprouts have indicated, most of the classical geopolitical theories were marred by the fact that they did not deal adequately with the time dimension, particularly with the impact of changing transportation technologies over time.[29] It would seem that it is precisely this dramatic increase in transportation capacities that has misled both the average man and the social scientist into believing that the spatial dimension, in either national or international politics, is one of negligible interest or importance.

For a culture that is so concerned with rapid change and "progress" (whether in effecting it or preventing it), it is curious that our attitudes regarding time are so unselfconscious. Few people, for example, are aware that time as we know it is really a standardized and abstract system imposed upon natural time—a point illustrated by the fact that twelve o'clock noon can be off as much as twelve minutes from the solar midday, depending on the season.[30] While astronomers and geologists concern themselves with determining the different time-spans over which various natural processes occur, social scientists seem to have devoted very little effort to such questions.[31]

[28]"One practical problem in integrating cultural and environmental data is that sociological statistics on which cultural indices are based refer to political units (counties, states) which frequently do not coincide with natural units (climatic zones, soil types, biomes, physiographic regions). As suggested in Chapter 2, the watershed is a practical ecosystem unit for management that combines natural and cultural attributes." Eugene P. Odum, *Fundamentals of Ecology*, 3rd ed. (Philadelphia: Saunders, 1971), p. 512.

[29]Harold and Margaret Sprout, *Towards a Politics of the Planet Earth* (New York: Van Nostrand Reinhold Co., 1971), pp. 268–297.

[30]Of course, people are more aware of the conventionalities of our calendar and some of its leap-year vagaries.

[31]Even natural scientists have tended to assume a constancy over time of many natural processes that it is now clear are governed by complex cycles and rhythms. See Lawrence E. Scheving, Franz Halberg, and John E. Pauly, eds., *Chronobiology* (Tokyo: Igaku Shoin, 1974), and Ritchie R. Ward, *The Living Clocks* (New York: Mentor Books, 1971), for descriptions of how research on cycles and rhythms of life forms may revolutionize biological and medical thinking. Even among physical phenomena, errors have been made by assuming constancy—such as those made in early carbon-14 dating because it was assumed incorrectly that ionization was constant over time.

Most social science is cast in the very comfortable terms of what I would call the *policy time-frame.* Discussion is in terms of the actions, decisions, and processes that we are all reasonably familiar with on a commonsense basis and that cover a maximum of a decade. Different disciplines and schools have of course staked out their own territories and hedged them about with complex analytic and verbal thickets. The complexity of modern society and the competing values of different approaches add to the above difficulties, but sensitivity to societal and environmental changes, common sense, and the specialists' inability to isolate themselves completely tend to be important correctives. Greater confusion appears when you consider the *developmental time-frame,* which encompasses roughly a century. Any conscious exploration of this time frame is found mostly in the work of historical sociologists, anthropologists, and economic historians (particularly Marxists). The confusion present in most theories of development seems to result from trying to think about changes occurring over a century in terms more appropriate to the decade span of policy making. The focus generally ends up being on policy measures and decision making rather than on basic structural elements like land tenure, the ownership of production, generational transformations, or population–resource trends, to mention only a few units of analysis more appropriate to a developmental time-frame.[32]

Even if you adopt units of analysis appropriate for studying developments over a century, the dangers of universalistic thinking must still be guarded against. The same unit of analysis—say, ownership of the means of production—will not be equally useful in studying different centuries. Nor will it even be equally useful in studying different cultures (and environments), within the same century. This problem has led to many of the difficulties of Marxism as it has evolved; it has been attractive to many analysts because it was perhaps the first formal mode of analysis to include many of the elements needed for studying processes and changes over a century, but it has also been attractive to dogmatists and revolutionaries. The result has been that its potential for flexible long-term analysis (which, it must be pointed

[32]It may be that the predilection of many social scientists to move back and forth between the academic world and the policy or policy-advising world tends to reinforce this tendency not only because of their primary concern with shorter-term policy measures and rewards, but because these larger topics are very difficult to deal with—particularly through policy or bureaucratic means.

out, is limited by its 19th-century faith in man's ability to control nature) has been largely vitiated by political factionalism, by attempts to apply it according to the political desires of the user, and by the search for functional equivalents to its often ill-fitting universal categories.[33]

The social sciences have devoted even less consideration to what can be called the *evolutionary time-frame.* In the past, there have been good reasons for this. There have been real theoretical difficulties in examining processes covering a number of centuries, not to mention the risks of speculative excesses such as those found among many Social Darwinists. On the practical side, up until the present century few evolutionary phenomena had much direct impact on the sorts of social and political phenomena social scientists are interested in. Today, however, the growth of technology, resource use, and institutional interdependence has made a number of evolutionary processes (like population growth and climate change) critically relevant to policy analysis. There has likewise been a vast increase in the global data base for the study of evolutionary phenomena. In trying to overcome the momentum of disciplinary specialization, social scientists will have to learn a great deal more about the physical sciences and biological evolution than has previously been common.[34]

It is important to remember that few theories in social science distinguish and discuss the use of different time-frames and their

[33]It would be most interesting to see some of Marx's categories analyzed in the sort of contextual manner that J. Peter Nettl uses in his most provocative article "The State as a Conceptual Variable," *World Politics* 20 (July 1968):559–592. He argues that the concept of the state, if broadly defined to include historical, intellectual, and cultural dimensions, is a useful comparative concept. He notes the great decline in interest in it, however: "Conceptual changes are both ideologically and geographically conditioned. The erosion of the concept of state coincides in time with, and is clearly a functional part of, the shift of the center of gravity of social science to the United States over the last thirty years, especially the acceleration in this shift in the last fifteen years" (p. 561). After reviewing the various historical and contextual aspects, he concludes that "to a considerable extent we may regard the problem of functional equivalence in the following terms: continental Europe—state; Britain—political parties; the United States—the law" (p. 586). That is to say that it is in these arenas that a relatively autonomous sector is found where political values are hammered out and created rather than simply represented.

[34]Two attempts at synthesis (each valuable in a different way) are: Peter A. Corning, "The Biological Basis of Behavior and Some Implications for Political Science," *World Politics* 23(April 1971):321–370; and Thomas L. Thorson, *Biopolitics* (New York: Holt, Rinehart & Winston, 1970).

correspondingly different units of analysis. This means that the benefits and limitations of each level of analysis will have to be more self-consciously estimated.[35] It's rather like maps: if you compare three of them drawn to different scales, say, 1:25,000, 1:250,000, and 1:2,500,000, you find the first gives admirable detail, but not the regional topographical features of the second. In the same way, the second map does not show the continental features of the third. Conversely, the scale increases at the cost of detail. The units of analysis (the features that can be meaningfully portrayed) are quite different for each scale. This analogy also helps suggest the role of the researcher's basic interests or values, because the scale of the map, the kind of projection, and the choice of subject matter (topography, soil type, land use, etc.) will ultimately depend on the personal and/or societal purposes for which the map is made.

A consideration of the problems involved in trying to map out agriculture and the specific place of the green revolution leads to the discovery that broader and better ways are needed to portray the various biological-social interactions, the different structural dimensions, and the position of the boundaries at any given time. Whatever the time scale, it is useful to include an analysis of the interactions between the natural environment, the man-made environment, and the psychological milieu.[36] This particular breakdown suggests that agriculture mediates between natural and man-made systems and is strongly influenced by local cultural beliefs and views of man's relationship to nature. These beliefs, which become embodied in the man-made environment (institutions and practices), act in turn upon the natural environment. No particular causal chain is postulated; rather, it is assumed that there are mutual feedback mechanisms in any particular cultural-ecological system that it is the analyst's task to trace out. Any large change or shift in any of the major elements will tend to

[35]See the important article by J. David Singer, "The Level-of-Analysis Problem in International Relations," in *The International System: Theoretical Essays,* edited by Klaus Knorr and Sidney Verba (Princton, N.J.: Princeton University Press, 1961). Even Singer, however, distinguishes his "levels" on a rather mixed basis, that is, one level is that of the states (a jurisdictional or organizational category), whereas the other level—international relations—is less clearly based.

[36]These are drawn, with some modification, from Harold and Margaret Sprout, "An Ecological Paradigm for the Study of International Politics " (Princeton, N.J.: Center of International Studies, Research Monograph No. 30, 1968). It must be stressed that these are strictly analytic distinctions and that the basic task of contextual analysis is to see how they all fit together into a whole.

redistribute the dynamic equilibrium of the system, rather like the redistributions that occurred when an ample Victorian lady was laced into her corset. A long-term change in environmental conditions (continuous drought, climate change, shifts in a river's course, etc.) will require major changes in agricultural practices, and perhaps in agriculture-related beliefs. A major cultural change, like a great religious awakening or revival, or like the shift from hunting-and-gathering societies to agricultural ones, will produce major shifts in agricultural practice and will undoubtedly have some significant impact on the natural environment. One of the basic questions of this study is what the effect of the shift from traditional styles of agricultural production (based primarily on solar energy and renewable resources) to industrial agricultural production (based largely on nonrenewable terrestrial energy) will be on the institutions and cultures of the non-Western world. We also have to ask what the long-term environmental costs and risks of such a shift are going to be—and if they are high, whether there are feasible alternatives.

The process of raising general questions that are then refined into specific research questions should be seen as related to, but not dependent on, the analytic approach chosen. The raising of questions (which is an art in itself) rests on the researcher's own basic sensitivity; however, I maintain (and hope to demonstrate) that it is valuable to have an analytic approach that explicitly recognizes and encourages the important role of such sensitivity. Such encouragement results, in part, because in using a contextual approach you must become more self-conscious about the kinds of questions you ask. The usual approach of asking general and abstract questions is only a preliminary to the hammering out of questions that fit the contours of a specific time-space context, for it is only the latter questions that have real meaning and significance.[37]

SUMMARY

In tracing the growth of theories, a number of points regarding their *relativity* have emerged. The large admixture of Western cultural values contained in theories that we like to think of as universal has

[37]One of the frustrations of writing a book suggesting a theory that stresses the need for modes of analysis that focus on specific contexts (at whatever level of abstraction) is that one is often writing in *general* terms.

been stressed. Culture is, of course, a changing, evolving thing itself, so that when we say that theories are relative to culture, we are also saying that they are relative to specific historical periods. The great differences between 19th- and 20th-century Western science have been blurred or obscured by the tendency of scientific writers to stress the essential unity, evolution, and traditions of science and by the quite different rates at which various new theories or discoveries have been assimilated by each specialized discipline. Historians and sociologists of science have only recently described the way in which scientific revolutions occur—with all of the conflicts that occur between ideas, between real men with various idiosyncracies, between generations, institutions, vested interests, etc.

Theories are also relative to their implicit or explicit time-frames, and an examination of them (and their corresponding units of analysis) goes a long way toward assessing how appropriate they are for examining various processes in the natural or social world. I have suggested that one of the major failings of the social sciences has been the use of a relatively short time-scale (the policy time-frame) to try to examine much longer-term processes, or alternatively to ignore them altogether. The complex ways in which theories are related to both individual researchers and general social milieus have been traced.

A final point that must be continually stressed is that theories are also relative to the subject matter studied—that is to say that they are to one degree or another true. Thus, the position taken here is not one of complete relativity. Rather, the clear implication of all of the above points is that it is only when one is in a position to estimate and discount all of the various cultural, temporal, and conceptual distortions found in Western science that one can get a more accurate estimate of the truth of a given theory or description. The problem is in doing it, and the major obstacle lies in trying to overcome strongly socialized beliefs that science and technology are both neutral and universal. Outlining the kinds of reevaluation, reinterpretation, and rethinking required to winnow out the Western dimensions of science and technology, particularly as they relate to agriculture, is one aim of this book. Another is to explore those alternative approaches to agriculture that flow from a contextual approach.

2

Historical Seedbeds

History celebrates the battle-fields whereon we meet our death, but scorns to speak of the ploughed fields whereby we thrive; it knows the names of the king's bastards, but cannot tell us the origin of wheat.
J. H. Fabre, *The Wonders of Instinct*

The preceding discussion of how theories are relative to specific cultures and periods, as well as to the realities they attempt to describe, is particularly relevant to any attempt to understand the historical dimensions of agriculture. It suggests that agriculture must be studied not only in terms of the various cultural and technological means which have been developed to crop particular ecosystems, but also in terms of the changing evolutionary parameters which give those ecosystems their distinctive (but also changing) characteristics. Perhaps this can be illustrated by referring to the diagram in Figure 2, developed by Ramón Margalef.[1] This illustrates clearly that when studying any important aspect of life, like agriculture, at three different historical points (A, B, and C), three qualitatively different configurations must be examined. Anthropologists do this implicitly when distinguishing between hunting-and-gathering and agricultural peoples, although they tend to account for the differences in terms of

[1]Ramón Margalef, *Perspectives in Ecological Theory* (Chicago: University of Chicago Press, 1968), p. 99

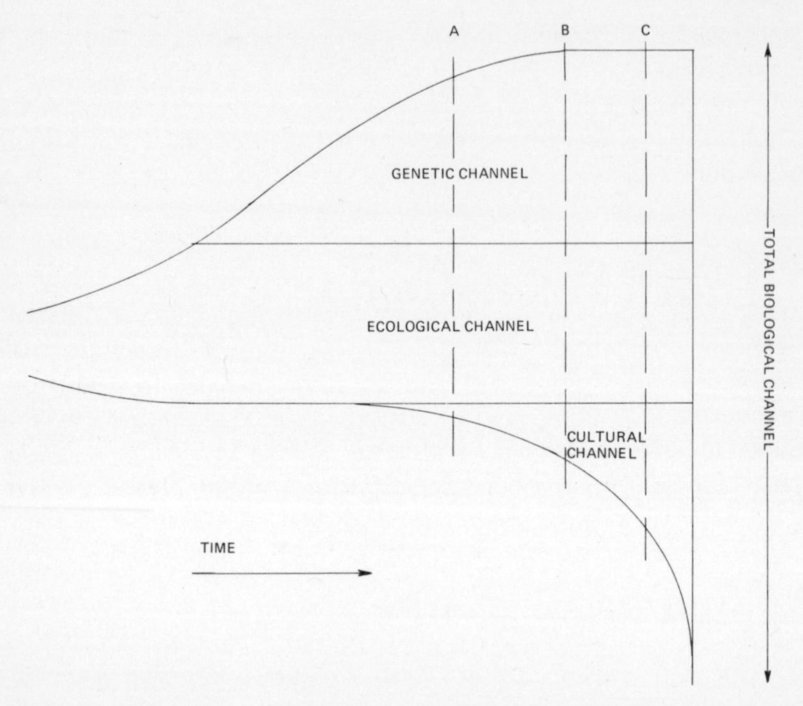

Figure 2. An idealized view of distribution along time of information forwarded by life, in three channels.

levels of technology rather than Margalef's rather broader concept of information forwarded by life. The value of explicitly recognizing the variability of major parameters (as well as the possibility of their nonexistence at some points) is that it encourages the search for, and the closer examination of, other variable parameters. For example, all the major elements of agriculture are variable when looked at in an evolutionary time-frame (see below). Unfortunately, there is a strong tendency for analysts using shorter time-frames to assume that "for all practical purposes" evolutionary trends and changes can be ignored, although this is somewhat less the case now than before. Population trends, potential climate changes, soil erosion rates, and growing global interdependence are receiving some attention, although rarely in an evolutionary sense. Any serious analysis of evolutionary trends highlights the weaknesses of simplistic and universalistic approaches to comparison of events from century to century. For example, changes in Egyptian land tenure made by the Romans in the second century are not directly comparable to the changes made in modern Egypt because the evolutionary parameters of each are significantly different, making the total configuration qualitatively different.[2] This is not to say that no comparisons are possible—rather that similarities and differences at the developmental level (between centuries) take on real meaning only if an analysis of the different evolutionary configurations is included.

Metaphorically perhaps we should speak of developmental plays taking place within the evolutionary theater. Not only are different plays put on in the "same" theater; over time the structure and capabilities of the theater itself change. It may be partially destroyed, or renovated, or there may be major additions. Through a sort of guided tour, let us examine the shifting "architectural" dimensions of the evolutionary theater wherein various agricultural plays have been acted out.

THE EVOLUTIONARY THEATER

One of the most critical and variable natural influences on agriculture is the weather. This applies even when you consider climatic changes occurring over long periods of time. From the extended time perspective of the geologist, man has always lived in a period of "revolutionary" climatic change.[3] In modern times, the following rather dramatic variations in climate have been noted:

> In the first century Europe was wet, and from 180–250 it was wet. The fifth century was very dry in Europe and Asia, and so dry in North America that many western American lakes dried up completely. In the seventh century Europe was so warm and dry that glaciers retreated permitting heavy traffic over Alpine passes now completely closed with ice. The study of tree rings shows that this period was also dry in the United States. In the ninth century precipitation became heavy, lakes rose. . . . The 10th and 11th centuries were again warm and dry. In fact, the Arctic ice cap may have disappeared entirely then. Greenland was settled in 984 and abandoned to

[2]For an example of the pseudo problems generated by trying to make such false comparisons, see Elias H. Tuma, *Twenty-six Centuries of Agrarian Reform* (Berkeley: University of California Press, 1965). Such attempts are typically made more often by social scientists than by historians.

[3]George H. Hepting, "Climate and Forest Diseases," in *Man's Impact on Terrestrial and Oceanic Ecosystems,* edited by William H. Mathews, Frederick E. Smith, and Edward D. Goldberg (Cambridge, Mass.: MIT Press, 1971), p. 214.

> the ice in 1410. Early in the 17th century Europe was very wet. The glaciers extended, and there were disastrous floods in northern Italy. The glaciers receded from about 1640 to 1770 and then advanced again until the mid-19th century. Since then they have receded to about their 16th century positions. The recent warming appears to . . . be a world-wide condition. . . . Through all the changes listed above there is little evidence of cycles. The type of weather and the duration of that weather seem to most investigators little short of random.[4]

Changes in climate affect not only precipitation and temperature, but through the hydrologic cycle, they have far-reaching effects on the distribution of vegetation and the production and erosion of soil. This critical evolutionary parameter would seem to be an obvious one to include in any major agricultural history, but the specialization and universalistic thinking of modern scholarship has led many commentators to blithely project recent climate patterns into the past as the norm.[5]

The various types of soil that man depends on for growing food and fodder are themselves the product of complex evolutionary processes; "Soil results from the meeting and fusion of the physical processes and the biological processes; it is the great bridge between the inanimate and the living."[6] On a global scale, there is a pretty fair correlation between climatic patterns, vegetation patterns, and soil types; however, the specific distribution of the soils is the result of the complex interaction over time of five factors:

1. Climate
2. The presence of various kinds of living matter, ranging from microorganisms to shrubs and trees
3. The structure and mineral content of the local parent rocks
4. The local relief pattern
5. The age of parent rocks[7]

The formation, location, maintenance, and erosion of soils can consequently be understood in the same sort of evolutionary terms as are applied to other living systems.

[4]Ibid., pp. 204–205.

[5]Agricultural economists seem particularly prone to do this. See, for example, Dwight H. Perkins, *Agricultural Development in China: 1368–1968* (Edinburgh: University Press in Edinburgh, 1969).

[6]Charles H. Kellogg, "Climate and Soil," in *Climate and Man: Yearbook of Agriculture* (Washington, D.C.: U.S. Dept of Agriculture, 1941), p. 266.

[7]Ibid., p. 270.

The long-term development of the plants and animals that man uses for food has been a process of increasing human intervention into natural evolutionary processes, beginning long before recorded history. By the beginning of the Neolithic period, agriculture was well established and most farm animals had been domesticated.[8] In succeeding centuries, the landscape was slowly transformed through gradual population increases and occasional innovations in agricultural technique.[9] The dramatic joining of the Old and New Worlds in the 15th century led to extensive transfers of grains and animals from one to the other (maize and the potato going from the New to the Old World, most livestock and poultry going the other way). However, it is only in this century that there has been an organized scientific effort to speed up and direct natural selection in animals and plants.[10] While most prior changes and transfers took place slowly and without too much difficulty—with the disastrous exception of the Irish potato famine—it is not clear whether the rapid and dramatic changes in the past few decades are sustainable:

> The species in natural communities fit together in complex ways that cannot easily be reconstructed when species are radically reassorted in intensely managed ecosystems. These coadaptive inter-relations among the species provide the stabilizing and sustaining processes of the ecosystem. Symbiotic interactions between seed plants and fungi and between algae and animals, mimicry among butterflies, protective coloration in

---

[8]Carl O. Sauer, *Agricultural Origins and Dispersals* (New York: American Geographical Society, 1952), p. 96. In evaluating the importance of prehistoric developments in agriculture, he notes that "Our civilisation still rests, and will continue to rest, on the discoveries made by peoples from the most part unknown to history. Historic man has added no plant or animal of major importance to the domesticated forms on which he depends. He has learned lately to explain a good part of the mechanisms of selection, but the arts thereof are immemorial and represent an achievement that merits our respect and attention." (pp. 103–104).

[9]For a model study of the complex historic interaction in England, see William G. Hoskins, *The Making of the English Landscape* (London: Hodder and Stoughton, 1955). A summary history of the impact of various techniques can be found in George E. Fussell, *Farming Technique from Prehistoric to Modern Times* (London: Pergamon Press, 1966).

[10]"We are well aware of the fact that long continued selection to meet the needs of man results in the loss of characteristics that are essential for survival in the wild. What is perhaps not generally recognized is the further consequence that the breeding system may be so altered—by the development of obligate self-fertilization or by the loss of sexual reproduction—that any further evolutionary changes may also be dependent on the active intervention of man." Joseph Hutchinson, "Crop Plant Evolution: A General Discussion," in *Essays on Crop Plant Evolution*, edited by Joseph Hutchinson (Cambridge: Cambridge University Press, 1965), pp. 174–175.

> insects, and the numerous defensive secretions of a wide variety of plants and animals attest to the ubiquity of such coadaptation. It remains to be learned to what degree man can disassemble these coadaptive complexes without incurring inordinate maintenance costs.[11]

Here, clearly, is a point fundamental to this study: any real evaluation of current activities and developments requires them to be understood in an evolutionary time-frame as well as in the more usual developmental and policy time-frames.

Man's evolution and his impact upon natural processes further complicates the attempt to understand and describe the evolution of the biosphere. As with other variables, care must be taken not to project current understandings of man back into the past as the historical "norm." There are certainly a great number of biological, genetic, and intellectual similarities between us and our ancestors;[12] however, there are also important genetic, physiological, and cultural differences that have resulted from different selective pressures in different environments and cultures. For example,

> Pastoral Maasai have been found to possess a unique biological trait—a high incidence of immunoglobulin IgA—that allows them to absorb the high cholesterol content of their high animal fat diet; this trait is thought to be genetically acquired and is identified as being directly related to (and probably adaptively acquired from) their practice of subsisting mainly on unpasteurized cow's milk—a practice that is likely to have existed for over a millennium in order for the trait to become genetically fixed among so vast a population.[13]

Even within the same cultural tradition, there can be important changes in behavior and perception from century to century:

> It is no wonder that the Europeans of early modern times were more "passionate" than we . . . the frustration and sublimation of the instincts were not so common as they are today. The men of the sixteenth century heard and touched more readily than they saw; the sense that for us has become primary—the sense of sight, which is central to science, to all ordering and classifying and rationalizing—for them came in third rank behind the senses of the ears and hands.[14]

[11]*Man in the Living Environment,* Report of the Workshop on Global Ecological Problems (Madison: University of Wisconsin Press, 1972), p. 100.

[12]For a discussion of these, see René Dubos, *So Human an Animal* (London: Rupert Hart-Davis, 1970), pp. 31–50.

[13]Alan H. Jacobs, "Maasai Pastoralism in Historical Perspective," in *Pastoralism in Tropical Africa,* edited by T. Monad (London: Oxford University Press, 1975), p. 409.

[14]H. Stuart Hughes, *History as Art and as Science* (New York: Harper & Row, 1964), pp. 35–36.

Even if you incorporate these points into your thinking, serious analytic problems remain in assessing man's impact upon the biosphere. Perhaps what is most important is how to assess the historically increasing technological capacity of man to influence his environment, which seems to have increased at an exponential rate parallel to the population increase. Some observers have emphasized the technological impact of man:

> It is man the geological agent, not so much as man the animal, that is too much under the influence of positive feedback, and, therefore, must be subjected to negative feedback. Nature, with our intelligent help, can cope with man's physiological needs and wastes, but she has no homeostatic mechanisms to cope with bulldozers, concrete, and the kind of agroindustrial air, water, and soil pollution that will be hard to contain as long as the human population itself remains out of control.[15]

Others, like Paul Ehrlich, have argued that it is not so much the kinds of organically indigestible technologies and wastes that man has developed that are the primary danger, but the sheer increase in his numbers.[16]

The controversy about the various ways in which man influences the environment has also led to renewed interest in one of the oldest questions in the Western intellectual tradition: How and to what degree does the natural environment of an individual or a culture influence or determine their character? The more deterministic views—starting with those expressed by Hippocrates in his essay on *Airs, Waters, and Places*—have had a rather checkered history.[17] Interest in and emphasis on environmental determinism have waxed and waned with the shifting of social and intellectual forces. In the early modern period, the Renaissance produced a revival of interest in classical writings that had been enshrouded by centuries of medieval church doctrine. Machiavelli introduced his materialistic theories of society. A shift then occurred with the Reformation and Catholic Counter-

[15]Eugene P. Odum, *Fundamentals of Ecology*, 3rd ed. (Philadelphia: Saunders, 1971), p. 36. This view is somewhat similar to that of Barry Commoner in *The Closing Circle* (New York: Knopf, 1971).

[16]Paul R. Ehrlich, *The Population Bomb* (New York: Ballantine, 1968).

[17]For a classic review of the various views and theories in this tradition, see Franklin Thomas, *The Environmental Basis of Society* (New York: Century, 1925). Better than many contemporary social scientists Thomas examines not only the substance of the various theories, but also the shifting intellectual climates and paradigms of each particular historical period.

Reformation as they once again drew attention to less material things; however, this religious revival was challenged before long, and on several grounds. The opening up of the New World and the tremendous diversity of cultures and habitats documented by eager explorers challenged the doctrine of special (and single) creation. This was especially damaging because at that time there was no theory of evolution or adaptation.[18] There was also the growth of both rationalism and the natural sciences, each of which produced a great deal of observation and speculation about the influence of the environment—geography in particular—on human affairs. Montesquieu's *The Spirit of Laws* perhaps epitomized these efforts. Eventually, the rising tide of Romanticism produced a strong counter to these views, with Hegel representing the high point of the reaction against physical interpretations of history. Soon, however, new discoveries in physics led to renewed interest in materialistic theories, while Charles Darwin's work in evolution gave environmental theories a new fashionability. Later, the enthusiastic excesses of the Social Darwinists led to a discrediting of biological and evolutionary analogies as applied to society and left the field to the behaviorists and the positivists, who have ever since promoted their materialistic and reductionist theories.

Challenges to the intellectual domination of these schools of thought have been hindered by the pervading influence of specialization upon both thought patterns and the sorts of disciplinary socialization discussed above.[19] While within each discipline there is something analogous to the "possibilist" approach in geography (where geographic conditions are seen not as determining the behavior of men or cultures but as setting the limits of what is possible for a given culture in a given habitat), most disciplines tend to emphasize a high degree of causality or determinism. This tends to maintain the integrity of the discipline and makes it appear to be the "master" discipline. The advocacy of multidisciplinary approaches tends to threaten both the intellectual and social integrity of the disciplines and their claims to priority, making any major shifts in direction slow and painful.

When one reviews the evolutionary dimensions of agriculture, three points that are not part of the conventional wisdom emerge. First, by looking closely at the basic elements of agriculture (climate, soil, plants, animals, and men), you see not only that there are complex

[18]Ibid., p. 23.

[19]See pp. 7–9, Chapter 1.

interaction and feedback mechanisms, but that there are also significant and qualitative changes over time. This means that in comparing real situations, past or present, it is not simply a matter of working out the proper percentages of each variable in each case; you have to be concerned with finding the specific configuration of often similar but qualitatively different elements. You are not comparing bags of mixed nuts, where only the proportion of each type of nut varies, but cakes, where different proportions of similar ingredients produce quite different results.

Second, not only is there significant variation in the physical parameters of agriculture, there is also variation in historical and cultural understandings of man's relationship with nature. This applies as much to the best scientific theories of any period as to folklore or "prescientific" knowledge. It is curious but understandable that both the sociology of knowledge of Karl Mannheim and the sociology of science of Robert Merton exempt the fundamentals of modern science from the sorts of analysis that they apply to other subjects and periods.[20] As argued before (Chapter 1), the relativity of even modern science must be included in our analysis if we are not to miss critical cross-cultural, historical, and climatic differences in agriculture.

Third, when one recognizes the historical variability and cultural relativity of what are usually considered to be the fixed parameters of agriculture (and other fields), it becomes clear that conventional historical and comparative methodologies are in need of revision. Any sort of comparison must be sure to include both the differences in the parameters between different periods or cultures and the system-transforming effect that those differences may have had. The difficulties involved are formidable, as can be seen in otherwise impressive attempts by historians like Arnold Toynbee and Louis Mumford to look at history from an evolutionary perspective.[21]

[20]For a detailed and interesting discussion of these problems as they relate to the work of Thomas Kuhn (a historian) and his influential work *Structure of Scientific Revolutions* (Chicago: University of Chicago Press, 1962), see Herminio Martins, "The Kuhnian 'Revolution' and Its Implications for Sociology," in *Imagination and Precision in the Social Sciences*, edited by Thomas J. Nossiter, Albert H. Hanson, and Stein Rokkan (London: Faber and Faber, 1972), pp. 13–58.

[21]See particularly Arnold Toynbee's latest revised and abridged version of *A Study of History* (in collaboration with Jane Caplan) (London: Oxford University Press in association with Thames and Hudson, 1972) and Louis Mumford, *The Myth of the Machine* (New York: Harcourt, Brace & World, 1966).

DEVELOPMENTAL PLAYS

To facilitate a review of the changes in agriculture over the past several centuries, three broad periods have been selected as separate "plays." Tracing out the story line of each calls for discussing the rather different materials, questions, actors, and settings of each period to show its distinctiveness. There are certain overlaps between them, and some of the same actors and scenery reappear, but the basic settings and leitmotivs change. Modern periods are chosen both as a preface to the various policy considerations that form the final part of this book and because of the greater availability of data.[22] The dates assigned to the three are 1492–1660, 1660–1838, and 1838–1972.

1492 – 1660

The discovery of the New World by the Europeans changed the basic context of economic and agricultural affairs. Most of the earlier diffusions of crops and animals had occurred before recorded history, as mentioned earlier; those occurring after the discovery of the New World were comparatively rapid and more easily traceable. In spite of their importance, and in spite of the fact that most of the countries involved in sponsoring the great voyages of discovery were predomi-

[22]This division is based primarily upon a concern with tracing changes in agricultural development generally and in agricultural relationships between the Old and New Worlds particularly. The focus is upon agricultural development and the discussion is meant to be illustrative. Other divisions are possible depending upon the primary focus of the researcher. For example, Immanuel Wallerstein, in his stimulating work *The Modern World System: Capitalist Agriculture and the Origins of the European World-Economy in the Sixteenth Century* (New York: Academic Press, 1974), proposes the following divisions: 1450–1640—origins of the European world-system; 1640–1815—consolidation of this system; 1815–1917—conversion of the European into a global world-economy; 1917–present—consolidation of the capitalist world-economy. His divisions (which are perfectly defensible given his interests) reflect what he feels to be the critical structural transformations (especially the macro division of labor) between his various "worlds." These are defined as reasonably autonomous economic systems that often encompass several political systems. His general approach, while certainly compatible with the idea of developmental plays, neglects at a theoretical level the larger evolutionary setting and variations identified above. Also, his emphasis on economic and political structures and relationships (in the broadest sense) possibly means that other developmental parameters, such as those relating to environment, infrastructures, and technology, may not be sufficiently integrated into his work. To fully judge, we will have to wait for the other three volumes in his projected series.

nantly agricultural at the time, the major motive of these countries (Portugal, Spain, England, and France) was to break the commercial monopolies of the prosperous Italian city-states and the Hanseatic League. To pursue their grandiose plans, the ambitious monarchs of these rising powers needed wealth beyond what their agricultural resources could supply. They hoped to find it through new trade routes to spice-rich Asia and through the capture of treasure in the New World.[23] Their success not only increased the brilliance of their courts but led to widespread agricultural transfers.

The variation in the diffusion of food and commercial plants, livestock, and poultry from region to region is not surprising, given the sorts of ecological, cultural, and technical differences that appear in any serious contextual analysis.[24] Africa appears to have been the most receptive to new crops, both ecologically and culturally. During the 16th century, maize, cassava, sweet potatoes, groundnuts, and tobacco were introduced from America and were so widely diffused that by the 19th century, when the interior was explored as part of the great imperialistic competition between France, England, Germany, and Belgium, there was no memory that these were introduced crops.[25] Tobacco, maize, sweet potatoes, and groundnuts were introduced into China, also in the 16th century, but even widespread knowledge of these crops did not result in their use until several centuries later.[26] The reasons for this lie in such things as unaccustomed taste, transportation problems, and (perhaps most important) the existence of long-established and highly developed systems of agriculture.[27] Europe, of course, benefited from the voyages of discovery through the introduction of maize and potatoes. The Americas benefited from the introduction of most kinds of livestock and poultry (the llama, the turkey, and an early species of chicken being the only domestic varieties).

[23]Parker T. Moon, *Imperialism and World Politics* (New York: Macmillan, 1926), pp. 9–10.

[24]One of the basic elements affecting diffusion is the summer day-length to which plants are adapted. Thus, it is generally easier to transfer plants East–West than to shift latitudes North–South. For a detailed discussion of this and other factors influencing transfer of plants, see Juan Papadakis, *Agricultural Potentialities of World Climates* (Buenos Aires: Libro de Edicion Argentina, 1970).

[25]Geoffrey B. Masefield, *A Short History of Agriculture in the British Colonies* (Oxford: Clarendon Press, 1950), pp. 20–21.

[26]Perkins, pp. 49–51.

[27]Masefield, p. 21.

By the middle of the 16th century, more complicated interactions could be noted. Various crops and animals were shifted from continent to continent, and intensive tropical agriculture began. Tobacco, the first crop to be so grown, gave intimations of things to come. Early success was followed by a glut in Europe in 1639; the planters on Saint Kitts agreed to not grow tobacco for a year, the first restriction of a commodity's supply.[28] Starting with tobacco, Great Britain assumed an increasingly important role: serving as a clearinghouse for the transshipment of tropical crops. Native Virginia tobaccos were replaced by superior West Indian varieties, which had been transferred to England and grown there before being shipped back to America. The tobacco glut had further consequences: planters rapidly shifted to growing sugar cane, which required a change away from small individual holdings to large estates and, most insidiously, to the demand for large amounts of slave labor. This demand was increased even more with the British conquest of Jamaica in 1655.

1660 – 1838

A rather different period had begun by 1660; amateur tropical agriculture was replaced by centrally planned colonial developments supported by mercantilist economists. Their theories grew out of a situation where the demands of royalty for gold, men, and ships were increasing at the same time that trade competition between European states—particularly in items produced at home—was also increasing. The competition was a result of both greater industrial production and the improved ability of sea and land transport to handle bulkier commerce. The mercantilists tried to get around this competition by developing exclusive and complementary trade relations with their colonies so that raw materials and noncompetitive colonial products could be obtained and manufactures could be sold without competition.[29] The establishment of such trade relations was facilitated by the doctrines of the Reformation permitting monarchs to dictate their subjects' religion and to convert—by force, if necessary—the inhabitants of their colonies.

[28] Ibid., p. 22.

[29] Mercantilist theories reached their legislative fruition in England in the Navigation Act of 1660, the Staple Act of 1663, and the Plantation Act of 1673.

The mercantile system led to some interesting relationships between agriculture and politics. By incorporating commercialized tropical agriculture into what now amounted to competing state trading monopolies, attempts to gain fertile new lands or seek out potential new crops took on strategic dimensions:

> The preponderant importance of the sugar trade and of the West Indian islands in colonial commerce and politics in the eighteenth century, even when compared with the North American colonies, is hard to realize today. The influence of the West India merchants and planters in London, as in Paris, was so great that they could affect government decisions and influence strategy. To acquire or to defend a few square miles of sugar land, regiments of country yokels were sent out from England and France, to die like flies of yellow fever . . . .[30]

This also helps explain why, after a century of private interest in botanic gardens, the Royal Botanic Gardens at Kew were established (1760), to be followed 20 years later by a governmental botanic garden at Saint Vincent. These gardens were clearinghouses for the transshipment of different potential crop species as well as places where selection and crossbreeding of different varieties could be carried out. The strategic way such matters were handled in the 18th century is probably most analogous to the attempts of 20th-century governments to secure and maintain oil supplies or to remain competitive in research and technology.

The Industrial Revolution produced a continually increasing demand for vegetable oils and fats to provide lubricants, illumination, and so on. This demand brought Africa into colonial agricultural trade for the first time. By the late 1830s, African shipments of palm oil were superseded by shipments of oilseeds, particularly groundnuts; the fact that they were processed in Europe rather than Africa had other important consequences:

> The residual parts of oilseeds processed in Great Britain contributed as oil cake and fertilizers very greatly to her agricultural economy, and played in particular a notable part in helping to maintain the fertility of the soil in these islands from the nineteenth century. This growing tendency of Britain to import, as it were, fertility at the expense of other countries first aroused a feeling of resentment in continental Europe. The great German agricultural chemist Liebig (1803–1873) wrote in a somewhat flamboyant passage and with reference chiefly to British imports of bones: "England is robbing all other countries of the condition of their fertility. Already in her

[30]Masefield, pp. 30–31.

> eagerness for bones she has turned up the battlefields of Leipzig, of Waterloo, and of the Crimea; already from the catacombs of Sicily she has carried away the skeletons of many successive generations. Annually she removes from the shores of other countries to her own the manurial equivalent of 3 millions and a half of men, whom she takes from us the means of supporting and squanders down her sewers to the sea. Like a vampire, she hangs upon the neck of Europe—nay of the entire world—sucks the heart-blood from nations without a thought of justice towards them, without a shadow of lasting advantage to herself."[31]

This suggests that conventional histories of imperialism—which tend to concentrate on economic transfers—need to be supplemented with a thorough accounting of the different kinds and amounts of *energy* transfers, as well as animal, seed, and technological transfers.

It was not only in Europe that processed and commercial fertilizers were linked to increasing industrial power:

> In Japan one can distinguish three separate stages in the development and use of fertilizer. The first involved the use of grasses cut by the farmers themselves. Second, in the Tokugawa period (1603–1867) Japanese peasants began to shift over to the use of various commercial fertilizers—principally dried fish, oil cakes, and night soil. Finally, in the early decades of the twentieth century, industrially produced chemical fertilizers came into extensive use.[32]

In China, where the second stage was reached by the early 16th century with the introduction of beancake fertilizer, little change took place until the 1960s. This seems to have been possible (in spite of slowly rising man–land ratios) because of an increased use of night soil and beancake as fertilizers and because the population grew slowly until the early 1950s.[33]

The four great mercantilist empires were shattered between 1763 and 1823, indicating the great political and economic transformations of the period. Much of the French empire was ceded to Great Britain in 1763, which in turn suffered a blow with the breakaway of the American colonies in 1776. Between 1810 and 1825 the Spanish were excluded from South America, while in 1822 Brazil separated from Portugal. This period of collapse also saw the emergence of various democratic regimes, which replaced or overthrew aristocratic ones. Democratization was part and parcel of the growth of the Industrial Revolution and the

[31]Ibid., p. 41.
[32]Perkins, p. 70.
[33]Ibid., pp. 71, 76–78.

rise of the middle classes. It encouraged and facilitated the assertion of the prime interest of the middle classes: increased freedom to pursue their commercial activities. The doctrines of laissez-faire, originally expounded by the French physiocrats, took root most firmly in Great Britain, where Adam Smith's *Wealth of Nations* found a receptive audience. In his half-prescriptive, half-descriptive book, he explained why colonies were uneconomic (though on occasion necessary for reasons of national defense) and why trade should be unfettered. His arguments dovetailed neatly with the interests of the new merchant princes. The commercial and industrial lead that England had, and was to enjoy for three-quarters of a century, meant that they wanted to expand trade into the former colonial empires of others as well as into the large European market. The doctrines of free trade made this expansion easier, and their soundness soon appeared to be confirmed as Great Britain's exports to America rose well above preindependence levels.[34]

1838 – 1972

By 1838 the four mercantilist empires were gone and the British Government had made two important decisions regarding its remaining possessions. First, after the 1837 rebellion in Canada, Lord Durham's investigative report recommended that the Canadian colonists be given responsible self-government, something granted by Parliament and later extended to the other "white settler" colonies. Second, 1838 saw the final emancipation of the British West Indies' slaves. This exposed the sugar growers there to somewhat increased

[34]It is undoubtedly not an accident that since then the champions of free trade have been those countries that have the strongest industrial and trade base. In ecological terms, Margalef (pp. 16–17) argues that among a wide number of linked subsystems, including agrarian communities and industrial societies, "the second subsystem experiences more predictable changes through time. In so doing it stores information better and is a more efficient information channel. The first subsystem is subject to a stronger energy flow and, in fact, the second system feeds on the surplus of such energy. It is a basic property of nature, from the point of view of cybernetics, that any exchange between two systems of different information content does not result in a partition or equalizing of the information, but increases the difference. The system with more accumulated information becomes still richer from the exchange. Broadly speaking, the same principle is valid for persons and human organizations. . . . Such relations are compounded in a hierarchical organization and are reflected at every level."

competition from Cuba, Brazil, and Puerto Rico, where slavery was still practiced. This fact became an element in the great Victorian campaign to free the high seas from the slave trade and to encourage emancipation in other countries. This sort of international interventionism—backed by a mixture of idealism and interest—has since become a steadily increasing phenomenon, although the results often have been less beneficial. Even in the case of emancipation, serious difficulties arose because of the great emphasis upon moral questions rather than upon the problems the freed slaves would face:

> The imperial legislators, who were so impelled by their consciences to vote for emancipation, do not seem to have felt any moral compulsion to provide technical assistance for the slaves in attaining the yeoman status which [Lord] Russell thought so desirable. It is interesting to speculate as to what would have been the effect on West Indian agriculture, then and now, if experimental farms and extension services had at that time been available to work out a suitable local system of subsistence farming.[35]

Of course, to have established experimental farms and extension services would have been to go against the hallowed doctrines of laissez-faire.

The tragic strength of these doctrines was most clearly demonstrated during the Irish potato famine. From 1700 to 1846 Ireland had gone from a grain-based to a potato-based agriculture. This early version of a green revolution had also enabled the population to increase from about 2 million to 8 million in the same period. The dangers of overdependence upon monoculture were demonstrated in four brutal years from 1845 to 1849. During this time, some 2 million died of starvation or associated diseases, and some 2 million were forced to emigrate, often to die in the so-called coffin ships that transported them to the United States or Canada. Just how or when the potato blight was transported to the potentially highly susceptible region of Ireland is not clear, although there appears to have been an outbreak in Germany in 1830, when there was a particularly wet summer.[36]

Even though the British government had no control over the basic causes of the famine, its actions during the famine certainly added to the death toll rather than reducing it. Aside from the faulty scientific

[35]Masefield, p. 45.

[36]Cecil Woodham-Smith, *The Great Hunger* (London: Hamish Hamilton, 1962), pp. 84–95.

advice that orthodox scientists gave to the government regarding the danger and treatment of the blight itself,[37] the main social and economic strategy was that there should be as little interference in market forces as possible and that the whole relief operation must be made as efficient as possible. The key figure in relief operations was Charles Trevelyan, who from his position in the Treasury was able (through the backing of a man of similar views, Lord Russell) to gain central control over all administrative measures taken in Ireland regarding the famine. He also played a large policy role. Trevelyan, who can be seen as the prototype of the modern bureaucrat (and who was later a force behind the founding of the British civil service), ensured that a minimum of grain was imported for relief. To do otherwise, he thought, would disrupt the local grain market—which was almost exclusively for export in any case—and encourage pauperism. The belated public works program that was finally established was so hedged about with surveys, conditions, and financial certifications that the meager administrative staff in Ireland was soon overwhelmed. In the succeeding years of the famine, grain imports were reduced, as were the public works programs, and the decision was made to rely upon the local poor houses. As these were financed from local land taxes—which naturally dropped drastically during the famine—they provided little relief. In fact, the poor houses, with their requirement of residence, often turned into contamination centers for dysentery and other diseases. In desperation, Trevelyan had to turn to soup kitchens (with their nutritionally disastrous rations), but he later came to see this as a great administrative achievement:

> The famine was stayed. . . . Organized armies, amounting altogether to some hundreds of thousands, had been rationed before; but neither ancient nor modern history can furnish a parallel to the fact that upwards of three millions of persons were fed every day in the neighbourhood of their own homes, by administrative arrangements emanating from and controlled by one central office.[38]

It is doubtful that Charles Trevelyan or Lord John Russell ever comprehended the tragic results of their policies and actions, for they were

[37]Ibid., pp. 45–47, 94–97, where it is described how orthodox scientists, including Dr. John Lindley, the first professor of botany at the University of London and editor of the *Horticultural Journal,* saw the blight simply as a result of "wet putrefaction." They argued against the "fungal theory," correctly advanced by M. J. Berkeley, since he could offer no proof that it was the cause.

[38]Ibid., pp. 295–296.

doubly blinded—not only by an ethnocentrism that led them to project British conditions and values onto a quite different Irish world, but by a faith that certain "natural" economic laws took precedence even over those governing agriculture and health. The most charitable thing that Cecil Woodham-Smith could say after reviewing the full depths of the famine was that

> these misfortunes were not part of a plan to destroy the Irish nation: they fell on the people because the government of Lord John Russell was afflicted with an extraordinary inability to foresee consequences. . . . Even the self-evident truth, that Ireland is not England, was not realized by the Government in Whitehall; the desolate, starving west was assumed to be served by snug grocers and prosperous merchants and to be a field of private enterprise; bankrupt squireens, living in jerry-built mansions, with rain dripping through the roof, became country gentry, and plans for sea transport were made as if the perilous harbours of the west coast were English ports. . . . Much of this obtuseness sprang from the fanatical faith of mid-nineteenth century British politicians in the economic doctrine of *laissez-faire*, no interference by government, no meddling with the operation of natural causes.[39]

The above aspects of the Irish potato famine have been discussed at some length because they illustrate a persistent problem in colonial agriculture: the tendency for the mother country to introduce those practices, crops, and institutions most familiar to it. There are also many similarities to the way in which contemporary agriculture and the green revolution are discussed, administered, and guided by a reigning—but increasingly questionable—economic theory.

The potato was not the only crop to suffer from practices of overintensive monoculture. In the last half of the 19th century, the leading varieties of sugar cane suffered from what was then called degeneration or running out.[40] Only increasing knowledge of plant genetics and the establishment of government experimental stations to apply that knowledge to sugar cane eased the situation. Coffee growers in Ceylon were less fortunate. A great boom in coffee prices dampened what little concern there was on the part of growers when a fungus-produced coffee-leaf disease appeared in 1868, even though the one local biological expert, Dr. Thwaites, Superintendent of the Royal Botanic Gardens at Peradeniya, predicted doom for the industry.[41] In spite of extensive new plantings, exports that had been

[39]Ibid., p. 410.
[40]Masefield, p. 50.
[41]Ibid., p. 59.

1,054,030 cwt. in 1870 fell to 179,254 cwt. in 1886. Dr. Thwaites did his best to develop alternative crops, first turning to chinchona (a source of quinine) and, when the market for it crashed, to what is still one of the major exports of the country—tea. It was experiences such as these—as well as the growing importance of Africa in general—that led the British to establish a number of botanic gardens in Africa in the latter part of the 19th century.

The full impact of Western colonialism on tropical agriculture has perhaps been best highlighted by Elizabeth Whitcombe in her comprehensive case study, *Agrarian Conditions in Northern India: The United Provinces under British Rule 1860–1900.*[42] She demonstrates how Western attitudes regarding public works, economic efficiency, legal modernity, and land tenure all combined to seriously distort both the natural and the social environments of India. Interest in promoting public works—irrigation canals being the most important of these—was prompted by the hope of ameliorating the human and fiscal costs of famines and by the vision of increased agricultural exports, especially cotton and wheat, to provide for the mills and workers of England. The Ganges Canal was begun in 1854, and by 1877–1878 it was irrigating nearly 1.5 million acres of land.[43] The "Mutiny" of 1857–1858 became the occasion for the successful attack of English radicals and supporters of free trade upon what they viewed as an anachronistic remnant of mercantilism—the East India Company. Direct rule by the Crown led them to expect vast opportunities for investment that would be safeguarded and promoted by an efficient British administration:

> The attractiveness of the prospects of development was enhanced by the fact that India appeared to lack very nearly everything which, according to contemporary capitalist criteria, was required in order to tap her vast wealth—or alternatively, to progress to a state of civilization. After the troubles of 1857, wrote Sir John Strachey, "ten thousand things were demanded which India has not got, but which it was felt must be provided. The country must be covered with railways and telegraphs, and roads and bridges. Irrigation canals must be made to preserve the people from starvation. Barracks must be built for a great European army. . . . In fact the whole paraphernalia of a great civilized administration, according to modern notions of what this means, had to be provided."[44]

---

[42]Berkeley: University of California Press, 1972.
[43]Ibid., p. 8.
[44]Ibid., pp. 62–63.

The results of trying to implement this vision were, however, rather different from what its promoters expected. The building of vast irrigation networks did increase production, but primarily of export crops, and this at the expense of the autumn crops of millets and pulses upon which most of the population depended for their food and fodder. This occurred, in part, because it was easier to irrigate export crops and, in part, because the Canal Department encouraged their growth by collecting canal fees right after their spring harvest.[45] The canals also had the economic effect of increasing the power and wealth of those already well off. The British left the construction and control of the minor channels and distributaries to the local rulers and landlords, and of course only those with money to invest in their own lands or to lend to others were in a position to take full advantage of the opportunities offered by irrigation.

The impact of the canals upon the natural environment was equally distorted. The regions that had traditionally been the best producers (having the best soil and drainage) benefited most from irrigation. Other, more marginal areas flourished for a few years and then suffered serious declines because of overcultivation or salinity. By 1891, a combination of "flush" irrigation and extensive subsoil infiltration of canal water had seriously affected some 4,000–5,000 square miles (approximately 3 million acres) with salinity[46]:

> As regards remedial action, however, the Canal Department's dilemma was acute: warned against the evils of continuing the unsound practice of flush irrigation, it found itself unable to recommend an alternative since the substitution of lifts, at a necessarily higher cost, would infringe the sound economic principle of widening the water market to its greatest feasible limits, upon which principle the budget of this highly important revenue-earning department was squarely based.[47]

Caught up in its own bureaucratic imperatives and operating within a general framework of laissez-faire, the Canal Department and its programs imposed a number of severe environmental, nutritional, and social costs upon the large masses of India while benefiting the already well-to-do, not to mention the entrepreneurs in England.

The results of large-scale road and railway construction were similar. The drainage problems created by canal irrigation were com-

[45]Ibid., p. 15.
[46]Ibid., p. 11.
[47]Ibid.

pounded by the addition of new embankments and barriers, which rarely included any drainage measures. The railways also led to serious deforestation of the countryside, since trees were used for both ties and fuel. Deforestation meant not only soil erosion but the risk of floods. As wood became more scarce and expensive, cattle dung came to be used for cooking (and even processes like brick baking), and this important source of fertilizer for the poor was lost.[48]

Legal and administrative reforms introduced in the name of modernity also took their toll. Not only were the various usury laws repealed in order to give full play to market forces, but gradually a complex series of reforms introduced "private property," as it was then understood in England, and a new civil court system to enforce its attendant rights. The overlapping of jurisdiction between the new civil courts and the earlier revenue courts and the increased powers of creditors led to a spate of litigation. In commenting upon the costs of this "great political and social error," one contemporary asked,

> "Why force men to run the gauntlet through both series? It tends to make the Government to be considered as a rapacious tax-gatherer, instead of a liberal landlord, which it really is; and to foster the growth of a host of native pettifogging attorneys, to devour, like white ants, the substance of the landholders of all classes and grade."[49]

Nineteenth-century attempts at modernization in northern India—the introduction of British economic, judicial, and administrative principles—led to a serious deterioration of the environment, severe social and political distortions, and increasing impoverishment of the large mass of the rural population. These unanticipated results were produced by the supreme faith of the British in the universality of their ideas of progress. Unfortunately, they regarded as natural laws ideas that can now be seen to be a mixture of Western and Victorian.[50]

By 1880, the relatively brief heyday of laissez-faire was over in fact if not in name, thanks to the far-reaching state interventions called forth by imperialist competition. A combination of a whole new complex of economic, political, and intellectual forces led to a frantic scramble for colonies, markets, and spheres of influence, particularly

[48]Ibid., pp. 93–94.
[49]Colonel William Sleeman, quoted in Whitcombe, p. 214.
[50]Naturally, the economic interests of middle-class Victorians inclined them to ignore the plight of the lower classes, whether at home or abroad.

in Africa. This scramble is somewhat comparable to the arms race today; the motives of national prestige and security, the complex of vested interests, the appeals to the public, the competitive aspects—all are similar. Economically, British trade superiority was threatened by the rising economic power of the United States and Germany as well as by their new tariff barriers. Changing industrial demand plus new railroad and steamship capacities meant that competition for colonies and their resources (rubber, oils, metals, and fertilizers) increased. The search for new markets and resources was aided by the growth of mass nationalism, which allowed imperialists and nationalists, as well as the politicians supporting them, to make new kinds of political appeals, aimed at various basic emotions. Their doctrines of preventive self-defense drew upon fear, the doctrine of surplus population upon self-preservation, national prestige upon gregariousness and self-aggrandizement, and the sort of "aggressive altruism" epitomized by the White Man's Burden upon pride.[51]

These doctrines and emotions infused explorers and entrepreneurs as well as the expanding number of civil servants who were sent to rule the vast new territories that had been acquired. The free-trade aspects of laissez-faire had been shattered by imperialism, but colonial administrators still generally felt there should be a minimum of state interference within each colony, essentially providing for law and order and the impartial administration of justice. Colonial agriculture administrators were perhaps a bit more tolerant because of previous interventions in the form of botanic gardens and experimental stations. By the 1920s, the palpable need for an updated moral justification for colonialism was met by Lord Lugard in *The Dual Mandate in British Tropical Africa* (1922). His concept can be seen in the following passage:

> Let it be admitted at the outset that European brains, capital and energy have not been, and never will be, expended in developing the resources of Africa from pure philanthropy; that Europe is in Africa for the mutual benefit of her own industrial classes, and of the native races in their progress to a higher plane; that the benefit can be made reciprocal, and that it is the aim and desire of civilised administration to fulfill this dual mandate.[52]

[51]Moon, p. 68.

[52]Quoted in Geoffrey B. Masefield, *A History of the Colonial Agricultural Service* (Oxford: Clarendon Press, 1972), pp. 66–67.

The major thrust of the British colonial agricultural officer's activities also shifted; by 1935 it had moved from a primary concern with plantation agriculture to a formal description of duties that read as follows: "To investigate native methods of agriculture and to discover what is useful in them; to stimulate the improvement of the indigenous methods of cultivation or the adoption of new methods. . . ."[53] Even at the end of the colonial period, British agricultural officers had a sustaining moral rationale, rather different from that at the beginning of the century but still including an element of Western ethnocentrism: "Above all, they were supported by a vision in which they saw themselves as missionaries of science bringing hope and prospects of progress to underprivileged peoples in an undeniably beneficial task."[54]

Over this period they gradually learned through experience that in tropical situations Western crops and farm practices had to be adapted, or even given up altogether. Interestingly, such an awareness was hampered by ethnocentrism as well as the ad hoc growth of distinct administrative offices (each with its own empire-building tendencies) responsible for colonial agriculture in different parts of the world.[55] This division of responsibility hindered the flow of information regarding tropical agriculture and, more important, restricted the transfer of experienced officers from, say, India to Africa. As a result, there were a number of false or partial starts in dealing with the problems of tropical agriculture.

The African progression went something like this. The first problem was that the experience gained previously in other high-rainfall colonies (which had perennial tree crops such as coffee, cocoa, tea, chinchona, and rubber, or sugar cane) fit neither the ecological nor the crop patterns in Africa, where both food and export crops were annuals. Also, there was little knowledge of appropriate rotation, manuring, or disease control practices, even though the traditional cultivators had used an ecologically sound approach in their shifting cultivation (albeit one permitting only a small population in relation to

[53]Masefield, *Agriculture in the British Colonies,* p. 70.

[54]Masefield, *Colonial Agricultural Service,* p. 6.

[55]Ibid., p. 4, lists the following divisions as of 1935: The Dominions Office, which handled not only the Dominions of that date, but South Africa and Southern Rhodesia; The India Office, which handled India and Burma; The Foreign Office, which dealt with the Anglo-Egyptian Sudan and the New Hebrides; and The Colonial Office, which ran the Colonial Agricultural Service in the "Colonial Empire."

the available land). While in any case this system could not have survived unchanged very far into the 20th century, it collapsed less because of rising populations than because of discouragement: "The first impulse of the European agricultural scientists who arrived on the scene in the New Agricultural Departments was to condemn this practice roundly, because it conflicted with everything that they had been taught."[56] After World War I, when the deterioration of soil fertility caused by European methods became apparent, the "remedy" suggested was based upon another British agricultural mainstay—farmyard manure. The benefits of this approach were rather localized because there were few cattle herds near farming areas, and where there were, the manuring gave good results only if it were a drier region. An attempt was then made at green manuring through the introduction of leguminous crops into the rotation pattern. In general, the results were disappointing except on the most favored soils. In the mid-1930s, experiments based on local observation showed that a rotation system including several years of planting the deep-rooted indigenous elephant grass gave good results. Later research showed that the reason for this was the ability of the grass to restore the crumb structure of the soil, which turned out to be the limiting factor—not the soil's nutrient content.[57] It was also discovered (in cotton cultivation) that continual weeding in the European style produced poorer results than little weeding, because continually disturbing the soil destroyed its structure and exposed its organic matter to erosion.[58] It was later found that less weeding, except at the early stages, proved beneficial in many parts of England.[59]

[56]Masefield, *Agriculture in the British Colonies,* p. 76.

[57]Ibid., p. 78. Masefield goes on to point out that "The conception of the resting period under grass is, of course, in a sense a return to the indigenous idea of shifting cultivation, though it is much less wasteful of land, because if fallowed plots are deliberately planted with grass to obtain a quick cover, the period required to restore fertility is much less than if the slower natural regeneration is left to take place. It also gained a ready acceptance because it happened to be in sympathy with a movement for 'alternate husbandry' or 'ley farming,' which at the same time, though for largely different scientific reasons, was being vigorously advocated in Britain by Sir George Stappledon and others, and in America and elsewhere" (p. 79).

[58]Ibid., p. 88.

[59]That the British Agricultural Service eventually overcame many of these early prejudices was shown by the large drop in rice yields in Malaya during the Japanese occupation. Not only did they for the first time bring in the dangerous blast disease by careless seed importation, but "they made exactly the same mistake as the British in early colonial days, of trying to introduce agricultural methods from the homelands into a tropical environment. . . ." (Ibid., p. 72).

As in India, it was not only agricultural practices imported from the West that had far-reaching negative impacts. In those parts of East Africa where Lord Lugard had had the private ownership of land written into the settlement agreements, soil exhaustion and erosion were often worse than in areas where tribal tenure prevailed. In fact, 40 years after these agreements,

> no local agricultural officer would have been prepared to defend the system of absentee landlordism into which it had grown. . . . At the same time no one dared to suggest its abolition, for it had become a vested interest and the people clung to it.[60]

The agricultural relations of the imperialist powers and the colonies were altered during World War I, when there was a frantic scramble among the European countries to boost domestic production; this desire for self-sufficiency was further promoted by the postwar economic boom. The Great Depression, which followed, was made worse for the many farmers affected, by the Dust Bowl in the United States and other less dramatic but extremely serious examples of soil erosion around the world. These events, the trade dislocations caused by the economic protectionism of the Depression, the gradual development of synthetic replacements for tropical products, and the trauma caused by World War II helped lead to the unexpectedly rapid rate of decolonialization after the war.

The colonies' euphoria at political independence soon gave way to the frustrations of economic dependence (the complexities of which are discussed later on). While much of this continues, it was clear by 1972—the year of the UN Conference on the Human Environment in Stockholm—that a number of basic patterns had changed sufficiently to mark the end of the old imperialist era. Since then, the earlier charges of neocolonialism have been muted and largely replaced by concerted efforts to achieve a "new international economic order." The confidence of the industrial countries in controlling resource supplies and traditional market structures has been shaken by the OPEC oil embargo. Environmental issues, once seen as a hobby of the industrial countries, are increasingly regarded by developing countries as an additional and significant means of attacking the imbalances between the rich and the poor countries.[61] Whether we are entering a period of

[60]Ibid., p. 115.

[61]See especially "The Cocoyoc Declaration," which is reprinted in full in *International Organization* 29 (Summer 1975): 893–901.

change comparable to the shifts between the three plays described here or a more profound one, comparable to the transition from feudalism to the modern state system, is unclear; in either case, the approach taken to agriculture will be of paramount importance. In the rest of the book, I will try to chart the implications of two different approaches. One is epitomized by the green revolution, a path that will be shown to be extremely risky in the long run. The other, visible only in outline, will require both a reconceptualization and many practical reforms but will be ecologically sound and will keep open our evolutionary options.

SUMMARY

Agriculture has been reviewed in the context of both its larger evolutionary theater and the various modern plays involving it. It was shown that over time the theater itself has changed in many significant regards: climate has changed; species of plants and animals have evolved and adapted to changes in the climate; both man and his basic attitudes regarding nature and the soil have changed—leading to changes in the style of cultivation; and man has developed technologies that have changed his relation both to the soil and to his fellow man. Within this changing theater, a number of rather different agricultural plays have been presented to rather different audiences. The reviews of the three most recent plays are just that—reviews. The text of those plays (the detailed history) has not been presented—less because of the lack of detailed background on the part of the reviewer than because most of the histories that have been written contain two fatal flaws: they assume that the theater is a constant that can be taken for granted (thus leaving out a number of critical dimensions and changes), and they assume that the different plays can be understood in terms of some sort of universal scientific/historical scheme. The fact that modern science and historiography contain large admixtures of Western beliefs and culture means that those who are unaware of both the temporal and the cultural relativity of the tools with which they work are bound to produce very *partial* histories of agriculture (in both senses of the word). The sort of contextual analysis that has been used here hopefully demonstrates the kinds of issues and questions that will have to be included in any ecologically, as well as socially and politically, sound history of agriculture.

3

The "New" Seeds and the Logic of Their Growth (Or Jack and the Beanstalk Revisited)

There was little logic in Jack's trading his mother's cow for five magic beans rather than selling it as she had instructed. Jack's story is a typical example of what folklorists call a "foolish bargain." Ultimately, in Jack's case, the bargain pays off handsomely, although many other such stories end in disaster. There are some rough parallels between the story about Jack and the story of the green revolution. In a sense, we are trading a faithful if not very productive source of food—traditional peasant agriculture—for a new and apparently more productive source based on "new" seeds with almost magical qualities. The question to be pursued in this chapter is whether we are engaging in a "foolish bargain" in making such a trade and, if so, with what consequences? The question is basic; unlike the story of Jack (and its predecessors), which involve a type of speculation regarding the origins of man and his separation from the heavens, we deal with matters relating to the future prospects of most, if not all, of mankind.[1]

[1]For a full discussion of "Jack and the Beanstalk" and its antecedents, see John A. MacCulloch, *The Childhood of Fiction* (New York: Dutton, 1905), pp. 432–449. He gives

RESUME OF THE DEVELOPMENT OF THE "NEW" SEEDS

Role of the Foundations

Unlike the story of Jack, where a mysterious little man offers the five magic beans to Jack in exchange for his cow, the development of the "new" seeds was a planned experiment on the part of the Rockefeller Foundation to see if the advances in plant breeding in the temperate zones, which had so increased wheat and maize production in the United States and Canada, could be transferred to semitropical zones. To find out, the Rockefeller Foundation sent a team of three American agricultural scientists to Mexico in 1941 to survey the prospects for increasing grain production there. The nature of the program that was established on the basis of their favorable report has had a large influence upon the subsequent evolution and elaboration of agricultural research for the developing countries and upon the kinds of agricultural, social, and environmental changes brought about by the green revolution.[2]

Part of the reason for the great influence of the Rockefeller program was its success in achieving, at least with wheat, its major objective: "increasing food supplies as quickly and directly as possible by means of the genetic and cultural improvement of the most impor-

---

numerous examples of early myths where heaven and earth are connected by a great tree (cf. the beanstalk) or a tower (cf. the Tower of Babel). In most versions, man gains something from the heavens—either by theft or by gifts—but in the process, the way to the upper land is lost forever. In the Dyak version, the hero is taught the secret of rice cultivation while in the upper land. In other versions, it is clear "that the gods are jealous lest men enter their abode, and share their immortality or their possessions—jealous, even, of men's prosperity" (p. 447).

[2]For their detailed history of this survey and the subsequent development of the Rockefeller Foundation program, "Toward the Conquest of Hunger," see Elvin C. Stakman, Richard Bradfield, and Paul Mangelsdorf, *Campaigns against Hunger* (Cambridge, Mass.: Harvard University Press, 1967). A very different and much more critical view of the Rockefeller program is contained in the important study by Cynthia Hewitt de Alcantara, *Modernizing Mexican Agriculture: Socioeconomic Implications of Technological Change 1940–70* (Geneva: United Nations Research Institute for Social Development, Report No. 76.5, 1976). This is one of a series of reports that form part of the excellent research project on "The Social and Economic Implications of Large-Scale Introduction of New Varieties of Foodgrain" conducted by UNRISD under the direction of Andrew Pearse.

tant food and feed crops. . . ."[3] The "demonstration effect" of the high-yielding varieties (HYVs) developed by the program influenced not only farmers, but also other foundations, national aid programs, and international organizations. While examining the organizational diffusion and momentum generated by the Rockefeller program, we need to trace the underlying direction of the program and some of the assumptions of the personnel involved.

Aided by some 30 years' experience, we can see by looking back that the program launched in 1941 included a major ecological insight: improved varieties for semitropical areas would have to be developed from the disease-resistant seed stock of that climatic zone rather than from varieties developed in the temperate zone. Much of the genetic success of the program rests on this insight. Little consideration was given to the geographic and cultural relativity of other parts of the package; as it turns out, to be most productive the new varieties require most of the trappings of modern industrial agriculture: irrigation, pesticides, fertilizers, large markets, extension services, credit, mechanization, and so on. The assumption of those working in the field was—and in many cases still is—that their job was to lead the way in transforming traditional agriculture into modern, "progressive" (read *industrial*) agriculture. Since those involved were biological researchers, there was little systematic awareness on their part of the great differences in agricultural structure from one developing country to the next in terms of land ownership, farm size, access to irrigation, distribution of wealth, and so on. They therefore failed to realize that in addition to the seeds, the whole Western approach to agriculture, with all of its built-in assumptions and physical requirements, needed to be adapted to the local conditions. The failure to have this larger holistic (and contextually specific) view from the beginning helps explain the continuing attempts to correct for the unanticipated, but predictable side effects of the green revolution.

[3]Stakman *et al.*, p. x. It was aided in this by a change in national priorities in Mexico. While the government of President Cardenas (1935–1941) had emphasized an agrarian socialism aimed at improving the life of millions of rural peasants, the new government of President Camacho sought rapid industrialization and quick increases in food production to feed rapidly expanding cities and to reduce the foreign exchange lost by importing wheat. Programs to aid medium-sized or large commercial farms were emphasized. These included not only plant-breeding research, but major infrastructural projects (new irrigation systems, rural electrification, new roads, and new rail lines). For a full description, see Hewitt de Alcantara, pp. 1–24.

The Rockefeller program faced two immediate obstacles: one was the lack of sufficient varieties of local grains to begin any sort of major breeding program; the other was the bureaucratic problem of trying to get the Mexican government to include enough Mexicans in the project to eventually make them self-sufficient in the distribution and use of the new seeds, once developed. The former was perhaps easier to deal with, since the collection and classification of maize, wheat, and bean varieties could be done by relatively small teams of experts who already knew the essential procedures involved.

The more difficult bureaucratic obstacles reflected not only inter- and intraministerial jealousies but the neglected state of Mexican agricultural education (at that time typical of many developing countries). In this regard, the general situation was likened to the agricultural position of the United States in 1883:

> Following the enactment of the Morrill Act in the United States, which became federal law in 1863, the various states established colleges of agriculture and mechanical arts. Then they groped around for almost a quarter of a century trying to find out how to be useful, because there was no coherent body of scientific knowledge for professors to teach and for students to learn. After trying this and that without conspicuous success, a few leaders in the United States realized that the colleges had to learn something useful through experimentation and research if their teaching was to be of any help. Therefore they urged the establishment of agricultural experiment stations to serve as living sources of progress in agriculture and agricultural education.[4]

In order to cut red tape and facilitate the establishment of its plant-breeding and experiment station, the Rockefeller team encouraged the setting up of a special high-level office in the Ministry of Agriculture. The general strategy was to use the new experiment station to give young Mexican *agronomos* internship experience. The best interns were then encouraged to seek degrees abroad (mostly in the United States) with the expectation that they would later aid in the modernization of Mexican agricultural education. One longer-term objective was to make Mexico self-sufficient in agricultural education and research; another was to develop mechanisms for disseminating the various

[4]Stakman *et al.*, p. 180. The analogy suggested demonstrates a lack of awareness of basic socioeconomic differences: the largely egalitarian distribution of land ownership in the 19th-century United States meant that the experiment stations served almost all farmers; in 20th-century Mexico, the new experiment stations served a small minority—those with enough land, water, and credit to benefit from the new seeds.

experimentation and research benefits as quickly and as widely as possible, both in Mexico and internationally.[5]

The leaders of the Rockefeller project viewed their role in almost missionarylike terms; they were dedicated to the project as "a school where the ideal was social service and accomplishment the badge of merit. The ideal was to provide more bread for hungry people, and the method for attaining it was intelligent dedication to the cause."[6] It can be argued that the very success of their specialized work in increasing grain yields led to a contempt for any barrier to the spread of their work as well as to a neglect of some of the other basic factors influencing agriculture in a semitropical developing country like Mexico.[7]

Many examples of this neglect flow from the curious failure to draw upon the extensive experience of the Europeans (and to a lesser extent the Japanese) in semitropical agricultural research, extension, and education.[8] In addition, it seems that no major effort was made in the early years of the project to estimate the impact of the quite different distribution of arable land, water, and infrastructural capacity between the United States and Mexico. The distribution in Mexico is

[5]This overall strategy of "starting from above" with agricultural research and working down slowly to extension work meant that "with the exception of a pilot project in the state of Mexico, no serious work was done on the problems of extension during the twenty years in which research was carried out under the direction of Rockefeller Foundation personnel, thus making diffusion of the new technology a haphazard affair" (Hewitt de Alcantara, p. 42).

[6]Stakman *et al.*, p. 282.

[7]Such contempt for political and bureaucratic obstacles can be seen in the following passage: "What are the alumni of the School for Wheat Apostles doing with what they learned? They have gone back to the countries from which they came, with ability and zeal to alleviate hunger by providing more bread and with hope for opportunity to do it. Some are realizing their hope; others have returned to turbulent or oppressive political climates that could have frustrated the most courageous. They went back home from a school where supercilious complacency, selfishness, self-aggrandizement, political manipulation, unnecessary bureaucratic restrictions and whatever else might obstruct progress in the fight against hunger were stigmatized as gross misdemeanors—not against scientists but against the hungry people who sorely need the services of the scientists to help them get more bread. Unless this concept is absorbed and applied by those who manage the affairs of retarded countries, their countries always will be retarded" (Ibid., pp. 282–283).

[8]There are several possible explanations for this attempt to "reinvent the wheel." In part, it probably related to American cultural beliefs (especially prominent at that time) that American know-how is superior. In part, it probably reflected the confidence inspired by victories in World War II and America's new global position. Finally, there was probably an aloofness carried over from America's lack of colonial involvement.

related to both wealth and culture, with Indian peasants being generally located on the smaller farms with poorer soils and scarcer water. There is also a regional factor because a high percentage of these small dry-land farms are located in the southern half of Mexico. Neglect of these factors led to strikingly different results between the wheat- and the maize-breeding programs, even though both employed similar research strategies:

> Both teams of scientists dedicated themselves to research programmes which were predicated on the eventual existence of an institutional environment . . . on the order of that to be found in the United States. Commercial wheat farmers, aided by numerous government programmes . . . were able to make maximum use of this new technology; but the greater majority of subsistence farmers who dedicated their land to corn and who were largely ignored in official investment plans, were not. Therefore twenty years after the initiation of the joint technical assistance programme in 1943, Mexican wheat yields were the highest in Latin America, while average yields of corn were among the lowest.[9]

Other factors that were initially neglected were rates of population growth, the impact of new varieties upon nutrition, and various questions relating to labor intensity, credit, and other administrative delivery systems. While the Rockefeller Foundation responded to these once they became obvious (see below), it is clear that some of the serious problems generated by the development of the HYVs could

[9]Hewitt de Alcantara, p. 26. This result evidently derived from the determination of the Rockefeller maize team to pursue strategy based upon hybrids rather than that suggested by scientists in the Ministry of Agriculture, who were concerned with developing varieties of use to the small peasant farmer: "There are several kinds of high yielding corn seeds. The highest yields are obtained with the so-called "hybrid" corns, but their exceptional productivity lasts only for the first planting. In subsequent plantings productivity declines so markedly as to give, at times, yields inferior to those which can be obtained with ordinary seeds—thus obliging the farmer to acquire new seed every year. A great number of the farmers who cultivate corn in Mexico cannot successfully use these hybrid corns, whether because of their limited economic resources or their limited education.

Open pollinated improved varieties are another kind of high yielding corn. Although they are somewhat less productive than the hybrids, they have the great advantage of permanence, and the farmer can use a part of his harvest for seed the following year exactly as our small cultivators are accustomed to do when they plant. Because of these characteristics, open pollinated varieties are better for our poorer farmers, if the latter can come to be almost as productive as the hybrids" (Hewitt de Alcantara, pp. 37–38).

have been reduced or eliminated if a more comprehensive, less ethnocentric, and less specialized approach had been pursued from the outset. In any case, the following is the order in which various major problems caused by the introduction of the HYVs received attention from those working on the Rockefeller projects.

It was not until 1949 that tests were made to determine the relative protein and tryptophan content of the improved bean varieties that had been distributed; tests showed that there had been a decline in both.[10] Since then, there have been moderately successful attempts to breed higher protein content into the HYVs, but there are problems with the somewhat bitter taste of most of the higher-protein varieties and with the fact that the protein is so located that much of it is lost in the typical small-size milling plant. An equally serious aspect of the protein–nutrition problem is that very little attention has been paid to the extent to which the more profitable HYVs have tended to replace fields of nutritionally valuable pulses. With the exception of Mexico, "data on changes in cereal and pulses production in selected countries which have utilized HYVs of cereals suggest that the expansion of cereal production might have occurred at the expense of pulses production."[11] In another area of nutritional concern, the Rockefeller Foundation funded a program in 1952 to try to increase the use of vegetables in the Mexican diet; in 1954 this research was expanded to explore methods of increasing potato production, a project undertaken by its then recently established research center in Colombia.[12]

It is not clear just when researchers began to notice that the HYVs were benefiting the already reasonably well-off farmers more than the poor ones. There is some hint of this in the 1953 decision to upgrade

[10]Stakman *et al.*, pp. 103–104. The authors claim this had no important effect on Mexican dietary standards (beans being a major staple among poorer peasants) because of the increase in production. This assertion is doubtful, however, because it assumes that the peasants also increased their consumption. The tests were made possible through the establishment in Mexico of a nutrition institute supported in part by the W. K. Kellogg Foundation.

[11]Ingrid Palmer, *Food and the New Agricultural Technology* (Geneva: United Nations Research Institute for Social Development, Report No. 72.9, 1972), p. 59.

[12]Here again, European experience appears to have been ignored, especially the work of Dr. Toxopeus, who collected some 200 types of potato in Chile and Bolivia right after World War II and established a seed bank with them at Wageningen, the Netherlands. (Communication from Egbert de Vries.)

and expand the small extension service operations of the Mexican project, the rationale being that the great majority of farmers are

> impoverished, illiterate, isolated, suspicious of strangers. Their farms are generally small, the soil poor and often worn out by centuries of mismanagement, the farming methods laborious, primitive and traditional. These farmers seldom travel any distance from home, so extension workers have to go to them.[13]

By 1968, it was obvious that extension work was useful in promoting the HYV package in irrigated areas, but that it did little to help the poor dry-land areas and that unless something was done, the already large gap between the two would increase. The Puebla Project was set up as "a large-scale experiment to discover whether it is possible to permanently increase corn yields under natural rainfall conditions, on small landholdings in cooperation with local farmers."[14] Other aspects of this project are discussed in Chapter 6, but it can be noted here that it represented a significant shift away from the original approach of trying to improve one major factor (seeds) and then developing various projects to foster its introduction. The Puebla Project involved more of a regional approach, attempting to increase production by all the farmers in a specific area. Interestingly, none of the hybrid maize varieties performed appreciably better than the local improved varieties under dry-land conditions; rather, production was increased through a combination of increased fertilizer use and the development of site-specific recommendations for cultivation that took into account soil morphology, planting dates, elevation above sea level, moisture availability, and so on.[15] One wonders what the shape of agricultural

[13]Stakman *et al.*, p. 207. Even so, most of the extension services went to better-off, irrigated farmers.

[14]*The Rockefeller Foundation: President's Review and Annual Report*, 1968, p. 9. As one observer has noted: "The social component of agricultural technology transfer is an issue to which the American agricultural technical assistance elite is slowly giving more theoretical attention. Apparently it took the publication of Theodore W. Schultz's *Transforming Traditional Agriculture* in 1964 to convince this elite that not all farmers are handcuffed by tradition. What Schultz said was hardly new. Yet his book became the "bible" of Sterling Wortman—a powerful member of this elite. Schultz's *Transforming Traditional Agriculture* also triggered the Puebla Project" [E. Vallianatos, *Fear in the Countryside: The Control of Agricultural Resources in the Poor Countries by Nonpeasant Elites* (Cambridge, Mass.: Ballinger, 1976), p. 64].

[15]Ralph W. Cummings, Jr., "Review of Plan Puebla," Rockefeller Foundation, 1974, p. 2. (Mimeographed.)

research in those developing countries influenced primarily by the Rockefeller approach would be today if Project Puebla had been started in 1941 instead of the original project for increasing production on irrigated soils through specialized seed technologies.

Expansion and Elaboration of Foundation Research Programs

So far the focus has been on the research aspects of the Rockefeller programs in Mexico and their attendant problems; now we need to look at the diffusion of their research "package" (both geographically and institutionally) as well as at its changing sponsorship. The following chronological description is illustrative[16]:

1946: One of the original hopes of the Rockefeller Foundation was met when its research program was made an integral part of the Mexican Secretariat of Agriculture.

1948: An additional goal was announced, that of cooperating with other Latin American countries in their agricultural programs.

1950: A research program was started in Colombia to study basic food crops, with the emphasis on maize and wheat.

1953: The beginning of the Central American Corn Improvement Project. Five Countries were involved: Costa Rica, El Salvador, Honduras, Nicaragua, and Panama.

1954: A research program was started in Chile on wheat and forage crops.

1954: A research program to increase production of the potato was added to the Colombian program.

1956: The beginning of a project in India to improve food crops (maize, sorghums, and millets) and agricultural education.

1959: The chartering of the International Rice Research Institute (IRRI) in the Philippines. The institute was to be financed jointly by the Rockefeller Foundation (providing the operating costs) and the Ford Foundation (providing the cost of construction and equipment). The institute was to be governed by a board of trustees made up of representatives of the government of the Philippines, the Ford and the Rockefeller Foundations, the University of the Philippines, and leading figures in the field of agriculture from various Asian countries.

[16]The chronology is drawn largely from the annual reports of the Rockefeller and Ford Foundations. Other sources will be indicated where appropriate.

1959: The establishment of the Inter-American Food Crop Improvement Program. This was patterned on the Central American Corn Improvement Project (1953) but extended the coverage both geographically and in terms of the different crops examined (wheat, maize, and potatoes).

1960: The establishment by the Ford Foundation of a $10.5 million program in India to provide an extensive education–demonstration program at the village level. Based on projections that there would be famine in India by 1966 if production was not boosted by 50% in five years, the program sought to develop in selected districts a "package" of inputs and practices to do just that.[17]

1963: The establishment of the International Maize and Wheat Improvement Center (CIMMYT) in Mexico with the backing of $1 million from the Rockefeller Foundation.

1967: Plans were made to upgrade the Colombian program to become the International Center for Tropical Agriculture (CIAT) to be jointly financed by the Rockefeller, the Ford, and the Kellogg Foundations.

1967: Plans were completed for the International Institute of Tropical Agriculture (IITA) to be located in Nigeria. It was to be jointly financed by the Rockefeller and the Ford Foundations.

1968: The establishment of the Puebla Project in Mexico by the Rockefeller Foundation to increase maize production on small dryland farms.

1969: The Ford Foundation called for United Nations or U.S. AID type of financing for the four international research institutes because of their high annual operating expenses ($12–14 million).[18]

1970: Money was budgeted by U.S. AID and the Canadian International Development Research Centre to help meet the operating costs of the four international research institutes.

1971: A shift in research emphasis at the International Rice Research Institute was announced to deal with rain-fed and upland soils,

[17]As the *Ford Foundation Annual Report* (1960, p. 85) described it: "On a scale never before attempted in India, all the requirements of increased farm production will be brought to bear when and where needed—adequate supplies of fertilizers and pesticides, improved seeds and farm tools, adequate farm credit, price incentives, efficient marketing, individual farm planning, and technical guidance."

[18]This call was contained in an essay by David E. Bell on the green revolution in the 1969 annual report (pp. 53–57).

to explore the problems of the small farmers, and to examine how to improve water management in rice cultivation and the protein content of rice.

1971: The start of a "modest" series of grants by the Ford Foundation to study malnutrition in Chile.

1971: The formation of the Consultative Group on International Agricultural Research (CGIAR). The consultative group is chaired by the World Bank, with the Food and Agriculture Organization (FAO) and the UN Development Program (UNDP) as co-sponsors. The group is unusual in that its members also include the UN Environment Program (UNEP), three regional development banks, three private foundations (Ford, Rockefeller, and Kellogg), the independent Canadian International Development Research Centre, the Commission of the European Communities, 18 donor countries, and 10 developing countries (2 chosen from each of the five major developing regions).[19]

1972: The establishment of the International Potato Center (CIP) in Peru.

1972: The establishment of the International Crops Research Institute for the Semi-Arid Tropics (ICRISAT) in India under the sponsorship of the CGIAR. Designed to be a "comprehensive" institute like CIAT and IITA, it focuses on dry-land farming and directs its research toward improving all major aspects of agricultural production systems, i.e., seeds, crop rotation, cropping patterns, cultivation practices, market relationships, etc.[20]

1974: Two initiatives were sponsored by the CGIAR to expand livestock research. One, the International Laboratory for Research on Animal Diseases (ILRAD) in Kenya, stresses disease research while the other, the International Livestock Center for Africa (ILCA), studies ways to improve livestock production.

[19]Briefing paper prepared for the World Food Conference by the Consultative Group on International Agricultural Research, Washington, D.C., March 1974, p. 1. (Mimeographed.) The group has appointed 13 experts to form a technical advisory committee to advise the group on research priorities. Annual contributions increased from $15 million in 1972 to about $65 million in 1976.

[20]These comprehensive institutes all express an interest in helping small farmers. However, the failure of the most famous of the small farmer projects—that of CIAT—raised basic questions about their ability to work directly with and help small farmers. While the CIAT project had internal problems among the team members, its potential social implications "frightened the CIAT board of trustees and the small farmer project was abolished" (Vallianatos, pp. 86–87).

1975: With the encouragement and financial support of the Rockefeller Foundation, the International Agricultural Development Service was set up. Its purpose is to provide management services for developing countries to help them set up longer-term development programs. It serves as a broker between the countries and the various funding agencies (private and public).

1976: Creation of the International Food Policy Research Institute. The institute grew out of a recommendation of the CGIAR. It is funded by the Ford and the Rockefeller Foundations plus the Canadian International Development Research Centre. Both of the above institutions can be seen as responses to the various problems highlighted at the 1974 World Food Conference (described in detail in Chapter 4).

This summary shows how the original Rockefeller initiative evolved in several different directions. At one level, the research package developed in Mexico became a kind of blueprint successively applied to try to increase crop production, first on a geographic basis (national, regional, hemispheric, and globally) then on a climatic basis (within the tropics, then the semiarid tropics). At another level, the product of this research—the HYV production package—was handled in a fairly typical technocratic manner; that is, once this new technology was developed, the major concern became how to transfer it to other parts of the world. All other aspects of agriculture—cropping, irrigation, and cultivation—were expected to change to meet the requirements of the new seeds. Finally, those social and economic relationships that inhibited the introduction of the new seeds were seen as impediments to be overcome.

Even so, lessons were learned as some of the more serious side effects of the green revolution became obvious; the newer centers moved away from a single-crop focus, sought to identify problems common to a particular climatic zone, and became more aware of the problems of the small farmer. Much remains to be done. Most CGIAR research still stresses increased crop production as the top priority and still limits its concern to what happens in the fields. This means that the study of basic relationships between agriculture, environment, rural development, and social and political structures has been largely neglected or left to others.[21] Whether it is possible to incorporate such

[21]One basic difficulty here is that those concerned about these problems rarely have either the range or the amount of resources available to do the kind of research and fieldwork that the CGIAR network does.

matters into the Rockefeller–CGIAR research approach or whether other "paths not taken" must be sought to do so is a basic question. Before it can be dealt with directly, the larger context within which the Rockefeller approach grew up needs to be examined.

National and International Aid Programs

Over the years, the efforts of various foundations to encourage food production in the developing countries were both aided and complicated by the general foreign aid priorities of governments, and especially by the complexities of food aid programs. National and international aid programs grew up in a post-World War II milieu of institutional optimism: the United Nations could prevent the outbreak of future wars, and foreign aid, which had proved so helpful in rebuilding Europe, would be equally efficacious in helping poorer countries develop. The outbreak of the Cold War dimmed optimism about the UN, and organizational efforts on all sides were directed toward regional defense arrangements. The hoped-for success of foreign aid and technical assistance in the developing countries was complicated by the upheavals of decolonialization; foreign aid became embroiled in Cold War attempts to secure allies, resources, and markets. The focus of such aid, especially United States aid, up until the watershed year of 1972, was on industrial "development" (buttressed by large amounts of military aid). The net effect of these factors was to put agricultural research and projects in a subordinate position, where the work done by AID, FAO, the World Bank, and others was seen as an adjunct to other, higher priorities. During this period, the only agriculture-related aid programs that had wide scope, a priority of their own, and general backing were the various food aid programs.

These national and international food aid programs had their origin in the seemingly endless grain surpluses that started to accumulate in the late 1940s, particularly in the United States and Canada. In 1954 legislation that blended the interests and concerns of farmers, shippers, bureaucrats, and humanitarians was worked out to distribute these surpluses abroad. Public Law 480 (generally known as the Food for Peace program) and its subsequent amendments have established two main forms of food distribution and assistance. Under what is now Title I, various commodities can be sold under concessional terms, either for local currencies or through low-interest, long-term credit arrangements. Title II provides commodities for humanitarian purposes and for economic development through sales at cost to voluntary relief agencies and through grants for disaster relief, for bilateral aid, and to the World Food Program (run by the FAO).

The mixture of motives involved in the establishment and continued operation of the Food for Peace program has been described as follows:

> In administering its food aid program, the United States is pulled in four directions. It must simultaneously strive to maximize the effectiveness of food aid as a political, humanitarian, and developmental tool; to protect its own commercial exports; to recognize and respond to the essential commercial interests of other nations as expressed through international agreements and bilateral representations; and to cope with internal pressures demanding that the food aid program be used as a lever to expand United States commercial exports.[22]

Each of these "directions" gets its impetus from one or more organized interest groups. P.L. 480 has also generated bureaucratic controversies between the U.S. Department of Agriculture (most concerned with protecting the interests of domestic producers) and the State Department and its affiliate, AID (most concerned with the foreign policy implications of food aid).[23]

The magnitude of United States food aid is demonstrated by the following figures. Between July 1, 1954, and December 31, 1970, the United States supplied $19.6 billion worth of food and fiber on concessional terms; the world total was $25 billion.[24] Nineteen percent of the U.S. total—$3.8 billion—was in the form of donations.[25] Since 1966, Food for Peace commodities have not been formally tied to surpluses; rather the Secretary of Agriculture certifies the availability of commodities. In any case, there is a high correlation between the types and amounts of commodities made available and the general surplus/price support picture, as can be seen in Table 1.[26]

Since the early 1970s, the distribution and amounts of United States food aid have changed significantly; we will come to that after a review of some of the major international programs.

[22]Robert L. Bard, *Food Aid and International Agricultural Trade* (Lexington, Mass.: Heath, 1972), p. 70.

[23]For a detailed discussion see ibid., pp. 22–24 and Peter A. Toma, *The Politics of Food for Peace* (Tucson: University of Arizona Press, 1967).

[24]Bard, pp. 16–17.

[25]Ibid., p. 20.

[26]1970 Annual Report on P.L. 480, Table 8, as given in Bard, p. 21.

Table 1. Distribution of Commodities under Food for Peace through 12/31/1970

Product	Value (millions)	Percentage of program
Wheat and wheat products	$9,824.2	50.1
Coarse grains	1,933.9	9.8
Rice	1,393.7	7.1
Fat and oils	1,736.6	8.8
Dairy products	1,628.8	8.3
Cotton	2,266.0	11.5
Tobacco and cigarettes	566.4	2.8
Others	230.3	0.1
Total	19,580.4	

The United Nations Children's Fund (UNICEF) was originally established in 1946 to provide emergency postwar supplies of food, clothing, and medicine to needy children. It has since become a permanent part of the United Nations, and its programs now focus on the developing countries. UNICEF provides for disease control measures, health education, nutritional programs and supplements, family and child welfare programs, and emergency aid—all directed toward the specific needs of children and their mothers. UNICEF programs are financed through governmental contributions, through private donations (half of which are collected in the United States and Canada on Halloween), and through the sale of cards and calendars. In its nutritional programs and emergency food aid, UNICEF has been hit recently with a combination of declining federal government contributions; increasing prices of commodities (particularly milk products), many of which are purchased through the Food for Peace program; and an increased politicization of the whole aid process.[27] The decreasing share of United States contributions can be seen in Table 2.[28]

[27]The *New York Times*, 23 May 1974, p. 13, reported that the United States had informed UNICEF that it did not want any of its contributions to be used for projects in North Vietnam or Communist-controlled areas of South Vietnam (in spite of the fact that UNICEF has had significant aid programs in South Vietnam since 1956). Such politicization also hit the attempts of the International Committee of the Red Cross to broaden the coverage of the 1949 Geneva Conventions when the North Vietnam and the Viet Cong delegations demanded that the Geneva Conventions should apply only to those parties fighting a "just war." For details, see Peter Haas, "A Humanitarian Crisis," *Swiss Review of World Affairs* 24(May 1974):25–26.

[28]Data drawn from the United Nations Yearbook (1970, 1971, and 1972) and information supplied by the U.S. Committee for UNICEF (*UNICEF REPORT* 1973, 1974, 1975, 1976).

Table 2. Sources of UNICEF Income, 1970 – 1976 (Millions of Dollars)

Year	Government total	U.S. contributions and their % of govt. total	Non-governmental	UN trust funds and other	Total income available for projects
1970	42	17.56 — 41%	14	3	59
1971	43	17.89 — 42%	16	5	64
1972	55	15.0 — 27%	18	8	81
1973	66	15.2 — 23%	18	12	96
1974	80	15.2 — 19%	19	16	115
1975	103	15.0 — 15%	22	16	141
1976 (est.)					145

The World Food Program (WFP), the major international food aid program, was established in 1963 as a joint project of the UN and the FAO. An earlier FAO pilot project had reduced the fears of some governments by showing that multilateral food aid would not cause displacement of commercial sales.[29] The United States gave strong backing to such an international program, as it wanted to encourage other developed countries to become more involved in providing food aid for the developing countries. Through April 30, 1971, the WFP had committed $725 million ($520 million in commodities and $205 million in cash) to some 492 development projects in 83 countries.[30] Pledging targets have increased in recent years:

1971 and 1972—$300 million
1973 and 1974—$340 million
1975 and 1976—$440 million
1977 and 1978—$750 million[31]

However, given the dramatic increase in commodity prices in recent years, these amounts actually deliver much less net food aid than before. During the early years of the WFP, the major policy questions were related to the amounts to be allocated for purely humanitarian

[29] David R. Wightman, "Food Aid and Economic Development," *International Conciliation* 567(March 1968):46.

[30] Bard, p. 26.

[31] For the first time in 1975 and 1976, final pledges significantly exceeded the target ($598 million being collected as compared to the $440 million target). Whether this will continue remains to be seen.

purposes and the amounts to go for development projects in the form of "food for work"—where food would be used as partial payment for labor-intensive public works projects. Also, during the period of surpluses, the exporting countries sought to keep food aid tightly controlled through a series of procedures called *usual marketing requirements* developed by the FAO Consultative Subcommittee on Surplus Disposal, a subcommittee of the Committee on Commodity Problems, which has traditionally been dominated by the developed agricultural exporters:

> These exporters' fear of the potential dangers of food aid to commercial trade has induced them to limit the jurisdiction of the Consultative Subcommittee from affecting the relationship of food aid to economic development. This deficiency has not been cured through the work of other relevant institutions. . . . This deficiency also characterizes bilateral, multilateral and international food aid efforts such as the World Food Program, the Food Convention [of the International Wheat Agreements], and other food aid programs related to commodity agreements.[32]

I mentioned that 1972 can be seen as a watershed: as the year of the UN Conference on the Human Environment, it symbolized recognition of new dimensions that would have to be included henceforth in international thinking and programs. The year 1972 was also one of important changes that have already modified and will continue to profoundly affect agriculture and food aid programs. What happened in 1972? In part, bad weather caused a global shortage of food grains. Drought in India led to a decline in production from 108 million metric tons in 1970–1971 to 95 million metric tons in 1971–1972.[33] Lack of protective winter snows led to the Soviet wheat shortage of 1973 and the resulting large purchases of wheat from the United States. Prices of agricultural inputs—especially fertilizers—began to rise. Given the dependence of industrial agriculture upon huge amounts of cheap energy for its fertilizers, fuel, transportation, packaging, refrigeration, etc., one would have expected some cautionary measures well before the oil embargo. However, since the availability of cheap energy has been one of those unquestioned underlying assumptions of both

[32]Bard, p. 287.

[33]Statement of James P. Grant (President, Overseas Development Council) to the Senate Subcommittee on Foreign Agricultural Policy of the Committee on Agriculture and Forestry, 4 April 1974, "Humanitarian Food Assistance in the New Era of Resource Scarcities," (Washington: Overseas Development Council, 1974), p. 4. (Mimeographed.)

theory (agricultural economics) and policy, the analysis of agriculture in terms of its energy efficiencies (and discussions of what that means for policy) has had to come from the outside, most of the work being done by natural scientists.[34] So far, these studies have had little impact upon the conventional wisdom or the agricultural establishment, and few corrective measures have been taken.

Overall, there have been serious shortages of production on a worldwide basis, large increases in the price of agricultural inputs like fertilizer, and more generally, increases in the price of energy. The balance-of-payments problems of oil (and fertilizer) importing countries has increased, leading the United States to search for larger cash markets for its agricultural produce (and weapons) to offset oil losses. Long-term contracts to sell grain to the USSR are an example of this. Such cash sales have naturally reduced the amounts of foodstuffs available for national and international food aid programs. In the case of P.L. 480, dollar aid as a percentage of total United States agricultural exports has gone down from 25% in 1960 to 16% in 1970 to only 5% in 1975.[35] The rise in grain prices has reduced the total food delivered even more. For example, 1974 concessional sales under Title I of P.L. 480 were only about one-third the tonnages delivered in 1972 (with similar appropriations in both years).[36] Both the quantities and the quality (high-protein foods) of the shipments under Title II—those for disaster relief, sale to voluntary relief agencies, pledges to the World Food Program, etc.—declined even more up through 1975 (see Table 3).[37]

In 1975, a number of amendments were made to P.L. 480. These resulted from concern about the decline of overall food aid and the way in which remaining food aid had been sent primarily to countries where the United States had national security interests. For example,

[34]For a comprehensive analysis showing that the United States food system (including production, distribution, packaging, refrigeration, and preparation) uses some 10 calories of energy for every 1 food calorie consumed, see John S. Steinhart and Carol E. Steinhart, "Energy Use in the United States Food System," *Science* 184(19 April 1974):307–316. See also, David Pimentel, William Dritschilo, John Krummel, and John Kutzman, "Energy and Land Constraints in Food Protein Production," *Science* 190(21 November 1975):754–761.

[35]Roger D. Hansen, ed., *The U.S. and World Development: Agenda for Action 1976* (New York: Praeger, 1976), p. 154.

[36]Grant, Attachment to Statement, p. 10.

[37]Ibid., p. 12.

Table 3. Food Shipments under P.L. 480, Title II, FY 1972–1975 (Thousand Metric Tons)

Commodity	1972	1973	1974 (est.)	1975 (USDA presentation)
Wheat and wheat products	1,614	1,649	718	628
Milk (dried)	115	26	0	0
Rice	248	33	0	0
Coarse grains	257	246	379	271
Soybean products	4	1	1	0
Vegetable oils	187	111	53	58

in 1974, half the wheat, two-thirds of the feed grains, and all the rice shipped under Title I went to four countries of high security interest: South Vietnam, Cambodia, Israel, and Jordan.[38] The amendments allow a maximum of 10% of P.L. 480 aid to any one country and require that two-thirds of P.L. 480 aid go to the 40 "most seriously affected countries" (the 40 poorest on a UN-compiled list).[39] In spite of both the increasing levels of United States food production and the increasing need for food aid—particularly in places like the drought-wracked Sahel region and in Asia—less food aid is being made available by the United States. As *The Times* (London) pointed out on March 29, 1974: "What the Americans finally decide will be crucial. They have been extraordinarily generous in their fat years, but now they are, to an extent, the 'Arabs' of much of the world's food supply."[40] Just as with oil and energy policy, what the United States decides to do in agriculture in the longer term has a great deal to do not only with a recognition of today's needs but with a more fundamental rethinking of long-unexamined basic premises. Many of these underlying assumptions —both explicit and unexamined—become obvious in any critical examination of the "difficulties" encountered when that package of seeds, fertilizers, and services called the green revolution is actually introduced on any scale in a specific locality.

[38]Ibid., p. 11.

[39]These amendments reflected the basic differences between the Nixon and Ford administrations and Congress on food aid (whether domestic or international). Similar differences were visible at the time of the World Food Conference, when congressional participants tried to increase the amount of U.S. pledges over administration targets.

[40]Quoted in Grant, p. 2.

THE LOGIC OF GROWTH

Agriculture ultimately gets down to one farmer planting seeds in the particular soil of a specific part of the earth. To get at the various factors making up the logic of agricultural development, the level of analysis has to be shifted from a concern with the international dimensions of HYV development and dispersal to a concern with the physical characteristics of the new varieties and the sorts of local agricultural, economic, and administrative requisites they have (as compared to traditional varieties).

Physical Requirements

The HYVs bred in Mexico are the product of specialists pursuing a very precise goal: the rapid increase of food crop production. It is not surprising that such a specialized product has a number of very particular requirements that relate directly to the assumptions made during their development. Given the tremendous variability of local weather and soil conditions in the semitropical areas, the best strategy at the time appeared to be that of utilizing the limited resources and specialized skills of those involved to develop a fairly standardized package of seeds and other inputs that would be relatively insensitive to soil and climate variations while still giving good yields. This package could then be "transferred" throughout these climatic zones with generally good yields resulting. As specialists in biological engineering who had little awareness of the important social, economic, and cultural factors that they were leaving out of consideration, the Rockefeller plant breeders undoubtedly gave little thought to what would have been a more ambitious, more expensive, but probably in the long term a more socially and ecologically sound approach: that of examining specific regions to find out what changes (in seeds, cultivation practices, nutrition programs, credit policies, water management, land titling, crop storage, or whatever) would most increase production and the general welfare of the rural population in that specific region.[41]

[41]In the original Rockefeller approach, an increase in welfare was simply assumed to result from increased production. This clearly has not happened and has led to one of the major conclusions and recommendations of the UNRISD research project: "It has already been suggested in earlier passages of this report that scientific research should be oriented as far as possible towards the capacities and interests of the small cul-

The package that did result includes the HYVs only as the most dramatic and visible element. Other necessary parts of the package are precisely those that are associated with industrial agriculture: irrigation, fertilizers, pesticides, increased mechanization, and a complex support and marketing system. The seeds themselves have been bred to be less sensitive to variations in day length, to be highly responsive to proper water management and fertilizer applications, and to have tough enough stems to carry heavy grain heads without lodging (falling over) in monsoon rain and winds.[42] The HYVs have, in addition, been bred to be earlier maturing (particularly rice, where IR–8 matures in 120 days as compared to 150–180 days for traditional varieties). This opens up more possibilities for double cropping rice or having a second, other crop. The new varieties of wheat have had their protein content increased from 9–12% to the order of 16–17%, but there has been less success in increasing the protein content of maize and rice.[43]

The other parts of the package—proper water management, fertilizers, and pesticides—are necessary to provide high yields, but just how much higher the yields from HYVs are compared to local varieties provided with the same inputs of water and fertilizers is not clear. Early estimates tended to be unrealistically high, either because they compared the HYV package to traditional varieties without extra inputs or because they neglected the fact that the HYVs were usually

---

tivator, taking account of input availabilities and costs, dietary needs and market demands, etc. *A social policy should be built into the new technology beginning with the basic research itself.* (Emphasis added)" [*The Social and Economic Implications of Large-Scale Introduction of New Varieties of Foodgrain: Summary of Conclusions of a Global Research Project* (Geneva: United Nations Research Institute for Social Development, Report No. 74.1, 1974), pp. 52–53].

Even in terms of crop production alone, there has been criticism that only the few originally successful approaches to plant breeding have been pursued, to the neglect of other promising approaches: "Varietal improvement through plant breeding should be extended into biologic processes that have received only minor attention. There is great potential for yield enhancement by selecting for heritable characteristics such as photosynthetic efficiency, nitrogent fixation, resistance to environmental stress, and selective ion uptake" [Proceedings of an international conference on *Crop Productivity-Research Imperatives* (Yellow Springs, Ohio: Charles F. Kettering Foundation, 1976), Book 2, p. 2].

[42]For details see Lester R. Brown, *Seeds of Change* (New York: Praeger, 1970), Chapter 2.

[43]Palmer, pp. 37–38. For a full description of the research being done on maize and the difficulties encountered, see the CIMMYT-Purdue International Symposium on *High-Quality Protein Maize* (Stroudsburg, Pa.: Dowden, Hutchinson, & Ross, 1975).

planted on the best soils. After a number of years of experience, it appears that *on average* the HYVs increase rice production some 25% and wheat production some 50%.[44] These figures vary considerably between seasons, between regions, and between different irrigation regimes. There is still a great need for more detailed studies that take into account the relative impact of variation in climate, soil, culture, and local ecology. The high cost of irrigation and the fertilizers that the HYVs have generally been bred to respond to raises questions about the best allocation of these resources. In many areas, fertilizer applications on HYVs have reached the point of diminishing returns; and some have argued that any additional fertilization ought to be used in the poorer regions, where (up to a point) the responsiveness of traditional varieties is greater.[45] This would help reduce somewhat the disparities between rich and poor regions (and farmers).

The physical characteristics and requisites of the HYVs are obviously far from neutral. They necessarily imply immense changes in traditional agricultural practices. To the degree that double cropping is adopted, this means a shift away from seasonal patterns of cultivation that have evolved over the centuries and that are normally accompanied by a variety of customs and rituals. This requires changes in both habit and belief—something difficult for any society (witness the difficulties in trying to get Americans to change their "traditional" habits and beliefs about energy consumption and large automobiles). To the degree that the whole package is adopted, it means planting the new varieties on irrigated land. In those many countries with large dry-land areas, attempts to improve significantly the already-favored position of those farmers on irrigated land risk a serious regional polarization as well as a number of intersectoral shifts largely detrimental to the small, dry-land peasant. Since the seeds require both fertilizer and pesticides (to protect the greater capital investment in-

[44]Dana G. Dalrymple, *Measuring the Green Revolution: The Impact of Research on Wheat and Rice Production* (Washington: USDA, Economic Research Service, 1975), p. 1, and Ingrid Palmer, *The New Rice in the Philippines* (Geneva: United Nations Research Institute for Social Development, Report, No. 75.2, 1975), pp. 81–90, where the large variations between regions that often are hidden in national averages are shown.

[45]Ingrid Palmer, *Science and Agricultural Production* (Geneva: United Nations Research Institute for Social Development, Report No. 72.8, 1972), pp. 30–31. As the prices of oil and fertilizer increase, this may become an even more important consideration.

volved), the farmers of developing countries will tend to become dependent upon external sources of supply (whether extraregional or foreign). There are also a series of economic and administrative requirements for the successful use of the seeds. These too result in far from equal distributions of benefits and costs.

Economic and Administrative Requirements

To appreciate the reasons for unequal distribution in benefits and costs, you must compare the economic and administrative requirements with the actual situation in most developing countries. The requirements themselves follow the same sort of logic that was used in developing the seeds: if you want maximum short-term increases in production of grains in the semitropics, then the new seeds must be bred to be less responsive to the natural variabilities of photoperiod, soil, climate, and rainfall and more responsive to inputs that can be controlled and supplied by man—irrigation, fertilizers, etc. The parallel logic runs as follows: if you want to introduce the new HYV package as quickly as possible, then the package must be introduced in areas where the necessary physical, economic, and administrative infrastructure exists or can be developed in short order. The HYV package was therefore introduced first in those areas where irrigation was already present because the presence of irrigation meant that there was also a fair level of infrastructural, economic, and administrative capability. The administrative requirements of the HYV package relate primarily to the delivery and timing of irrigation water and to the need to regulate various aspects of grain marketing. Since higher levels of production imply larger regional markets rather than small local markets, ways of grading the grain, supervising weights and measures, etc., need to be developed. The economic requirements of the package relate to the higher costs of inputs. Since only the largest farmers can obtain the extra capital or credit necessary to buy the new seeds, additional fertilizers, pesticides, tubewells, tractors, and so on, new types of credit mechanisms for middle- and small-size farmers are needed. Arguments have developed over the degree to which these should be wholly governmental, regulated by government, or left to the private sector (where the suppliers of these new items can be expected to be interested in providing credit). Whatever approach is favored, it is clear that resolution of the issue involves a number of vested interests and quite a bit of politicking. Other important economic and political issues relate to whether subsidies should be given to encourage the use of the HYV package, whether crop prices should be regulated or subsidized to favor production and good profits for the farmers (at the cost of high prices for the urban masses), or whether attempts should be made to ensure cheap food for the urban masses at the cost of low income for most farmers.

You are soon led to agricultural policy issues very similar to those in the industrialized countries. This is not at all surprising when you realize that the HYV package represents the core of industrial agriculture and that it cannot be successfully transplanted without also transplanting most of its supporting elements. It took the "development community" some 20 hard years of experience and frustration to learn that industry could not be rapidly exported, for exactly the same reason: in the long run, one integral part of an economic-societal-governmental package cannot be transferred without transferring most of the rest of it. The question remains whether it will take another 20 years to learn the same lesson about the HYV package.

As was pointed out in Chapter 1, our (Western) analytic models and theories have varying degrees of cultural and environmental relativity, depending upon the discipline in question. Knowing this makes it difficult to be very hopeful about the chances of shifting our agricultural thinking from the rather ethnocentric way in which most economists discuss the probable impact of agricultural innovations upon the developing world to a more empirically based approach. The conventional approach—that of the classical and neoclassical economists—is ethnocentric because it represents primarily an empirical summary of Western economic beliefs and practices. When it is applied to non-Western contexts, it becomes something that should be examined, tested, and proved or disproved (i.e., propositional). Since a great many economists tend to see their principles as "universal," their tendency is to ignore what may be fundamentally different social, political, or temporal factors and interrelationships in the non-Western context and to focus on such "pure" economic data as prices, wages, interest rates, and unemployment statistics. To the degree that social and political forces are recognized, they are seen as being irrational or distorting of "natural" market forces.[46]

A good example of a more empirically based and less ethnocentric approach is Keith Griffin's *The Green Revolution: An Economic Analysis.*[47] Rather than positing agricultural markets as almost "textbook pure," as many economists do, he surveys the functioning of rural agricultural markets in monsoon Asia and describes the various social and political dimensions that strongly influence the way these markets function in practice. He also examines the access to and the variability in the factors of agricultural production: land, labor, and capital/credit. He points out that in many regions, there is no market for land: it is transferred by inheritance or gift; also, the distribution and size of land holdings are shown to be correlated with political and social power. Of note is the tremendous range of interest rates paid by different borrowers.[48] Finally, he analyzes wage rates as they relate to landlord–tenant relationships. With an overview of the actual functioning of these agriculture "markets," including the great influence that large landowners and wealthy merchants and moneylenders can exercise over their operation, he is able to use conventional economic

[46]While conventional Marxist economists are more sensitive to social and political forces, there is also a fair degree of ethnocentrism in their approach since they also apply an essentially Western model to non-Western situations without examining whether the model needs to be altered in any fundamental sense to deal with the difference in context. For a critique of the urban bias of radical Marxists shown by their ignoring the green revolution, see Harry M. Cleaver, Jr. "The Contradictions of the Green Revolution," *Monthly Review* 24(June 1972):80–111.

[47]Geneva: United Nations Research Institute for Social Development, Report No. 72.6, 1972. See also Griffin's book *The Political Economy of Agrarian Change* (Cambridge, Mass.: Harvard University Press, 1974).

[48]Ibid., p. 22, where data (Table 1.5) from 224 rice farmers in the Philippines showed the following:

Reported interest rate	Percentage of borrowers
0	20.0
1–14	13.0
15–29	9.0
30–100	23.0
100–199	15.0

Later it was found that those reporting zero interest were actually paying 16% because merchants underpriced harvest repayments and overpriced goods purchased by the borrowers.

analysis to show that the marginal utilities of the small and the big farmer are very different when it comes to adopting a new technology like the HYV package and that one can clearly predict that the bigger, wealthier farmers will be those who both adopt the new technology first and benefit the most from it.[49] He thus concludes that the HYV package is "landlord biased" and that because of their connections with the urban areas, many of the economic gains are transferred from the rural regions to the urban centers—where unfortunately a very high percentage of the gains is used to support higher levels of consumption on the part of the new elites and civil servants.[50] Among other things, he argues for "peasant-biased" seeds (for dry-land farming) as well as for other measures that would reduce the ability of the large landlords and wealthy merchants to dominate the actual functioning of rural agriculture. The conclusions that Griffin comes to are quite different than those of conventional economists. The reason is not that he uses different econonic tools to analyze the green revolution, but that he first attempts to describe the actual functioning of agricultural markets as they are integrated into and reflect local distributions of political, economic, and social power.

[49]Griffin's analysis has been corroborated by a careful political study of the green revolution in India. As Francine R. Frankel points out in her book *India's Green Revolution: Economic Gains and Political Costs* (Princeton, N.J.: Princeton University Press, 1971), p. 206; "The majority of farmers—probably as many as 75 to 80 percent in the rice belt—have experienced a relative decline in their economic positions. Some proportion, representing unprotected tenants cultivating under oral lease, has suffered an absolute deterioration in living standards." There are a number of significant social and political implications flowing from this situation: "In sum, therefore, the rapid progress of agricultural modernization tends to undermine traditional norms of agrarian relationships based on the exchange of mutual, if noncomparable, benefits and services that have historically provided a justification for inequalities between the propertied upper and middle castes, and the landless low castes and Harijans. As traditional landowning patrons increase their advantages by striking margins, yet neglect to fulfill their previous function of providing security to client groups, the legitimacy of existing—and growing—disparities is increasingly called into question. The potential impact on rural stability is all the more serious because radical parties openly proclaim their intention of transforming social tension into political conflict between the minority of prosperous landowners and the large numbers of sharecroppers and landless laborers" (pp. 198–199).

[50]Griffin, pp. 103–105.

Levels of Logic and Learning

Just as seeds may flourish in one climate and not in another, some ideas and innovations flourish more in some societies, institutions, and disciplines than in others. Figure 3 attempts to illustrate the general pattern of institutional activity and response to the HYV package, depicting both the different levels and the various rates of perception or learning, depending upon the problem.[51] It suggests that the various institutions responded very quickly to the increases of production, while it took them much longer to become aware of and respond to problems of maldistribution of economic benefits and of unemployment. This is not at all surprising, as the people involved tended to have complementary configurations of cultural and disciplinary values. The professional plant breeder naturally tended to be most responsive to exactly those results that they were trying to produce. The various planners and development advisers involved generally tended to have the conventional economic orientation described above (whether they were Westerners or Western-trained) and consequently focused first upon the already fairly developed and irrigated sectors of agriculture in the developing countries. As a result, the impact of the HYV package upon both intra- and interregional (and sectoral) income distribution, upon unemployment patterns, and upon dietary and nutritional changes was of no great concern until the problems became so obvious that even casual observers commented upon them. Even so, for some time there was a tendency to argue that such problems were a painful but necessary part of the process of making agricultural production and markets more productive, modern, and rational; that is, the small, inefficient farmer had to pay the price of general agricultural modernization. Once it became clear that there was not sufficient industrial growth to provide jobs for those squeezed off the land, there was a shift in attitude.

Figure 3 also is meant to suggest that there are institutional variations in both the length of time it takes to respond to a perceived problem and the type of response. The institutional response varies according to the level of its operations, its dominant paradigm, and the

[51]The plotting of events on the time scale is not meant to be precise, rather it summarizes the general trends and rates that appear to have been present.

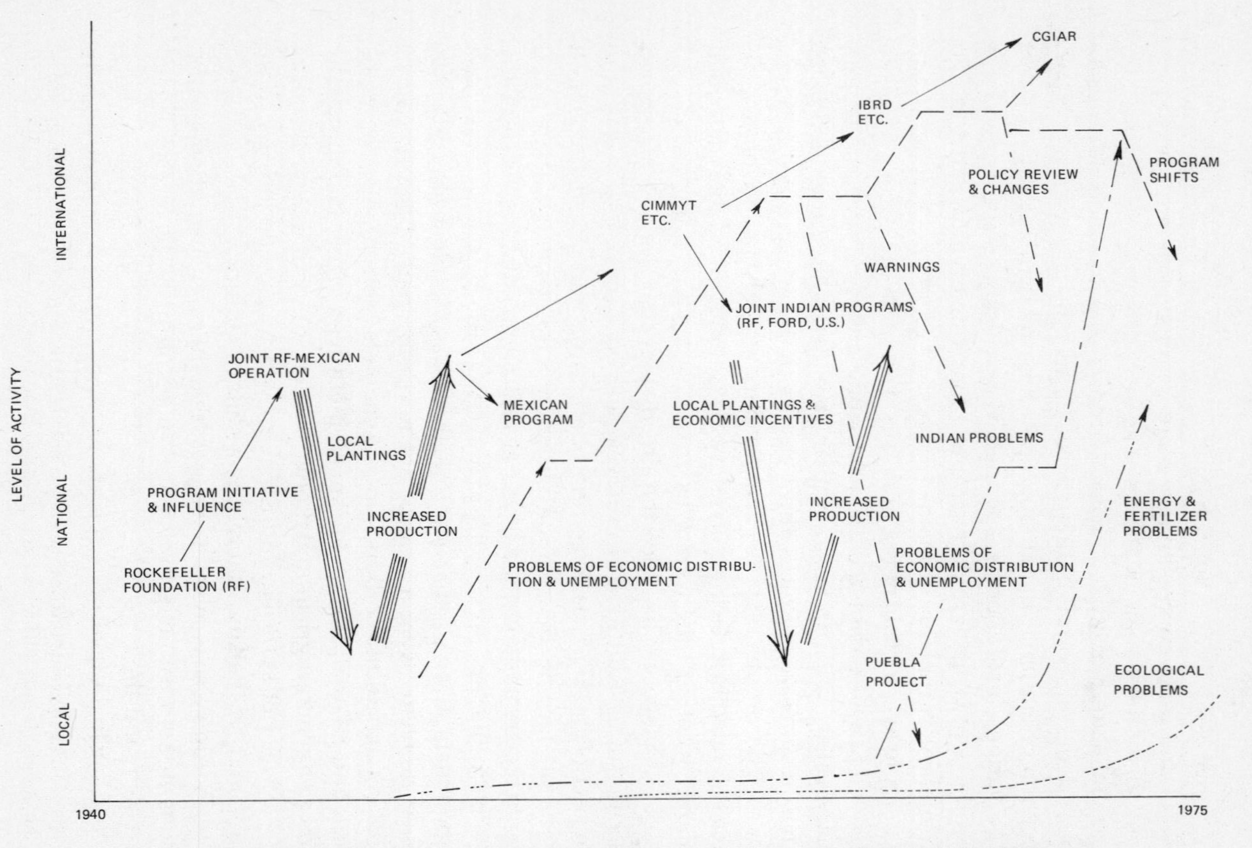

Figure 3. Summary plotting of institutional responses to production and problems related to HYVs.

momentum of its standard operating procedures.[52] As a private foundation with a fairly long time-horizon and a strongly professed concern for the welfare of people, the Rockefeller Foundation has been able to stay somewhat more sensitive to the human and ecological impact of the HYV package than other organizations at or above the national level.[53] As an institution dominated by bankers and their conceptions of fiscal responsibility, the World Bank was very slow to recognize the role of agriculture in development and to support the HYV package. An additional complication found at the international level is that the multilateral procedures involved in policy making and program implementation are time-consuming and permit various groups to slow down or block changes.

Most important at the international level has been the West's dominance over general conceptions of development and its control of the economic resources to further those particular conceptions. While there have been serious disagreements between rich and poor countries over terms of trade, tariff and quota barriers, and the division of the fruits of development, there have been few alternative conceptions of development put forth by the developing countries.[54] During the Stockholm Conference, the resistance of the developing countries to any inference by environmentalists that they should slow down or alter the modernization of their countries indicated that the elites in those countries still gave pride of place to current Western conceptions of development. The response of environmentalists (something confirmed by the impact of the OPEC oil boycott) was that there are longer-term ecological and energy considerations that tend to be fil-

[52]For a good summary of how these organizational-bureaucratic influences operate, see Graham T. Allison, *The Essence of Decision* (Boston: Little, Brown, 1971).

[53]While the question has not been researched here, it is doubtful that the ministries of agriculture in the affected countries (Mexico, India, the Philippines, etc.) were more sensitive to these problems. This is because of their general political links with the larger, wealthier farmers. Even if they had been more sensitive, their policy recommendations for easing these problems would generally have been overruled by more powerful ministries (i.e., finance, economics, and planning).

[54]Ghandi's conception seems to have been lost in the past 20 years—insofar as it actually has been applied to India. Part of the difficulty here is that most of the alternative conceptions that have had any success are fairly culturally specific and thus are neither easily presented nor transferred. For an interesting discussion of some of these attempts, see René Dumont, *Socialisms and Development* (New York: Praeger, 1973), which includes, among others, analysis of the experiments in Zambia, Tanzania, and Algeria.

tered out by those with more conventional economic orientations. While Figure 3 shows these as exponential trends, it is not clear whether they are in fact exponential or whether they end up being exponential in their impact because they have been ignored for so long. To deal with this type of problem and to try to weigh some of the longer-term costs, benefits, and risks of the green revolution, we need to refer again to the three time-frames used in Chapters 1 and 2.

THE COSTS, BENEFITS, AND RISKS OF THE GREEN REVOLUTION

Assessing the costs, benefits, and risks of anything leads to the question: What standards of judgment are to be used to make the evaluations? In addition to drawing upon the general theoretical perspective outlined in the first two chapters, the following assessment of the green revolution is based on an alternative and ecologically based model of agricultural development that will be made explicit as we go on.

The Policy Time-Frame

Many of the successes, problems, and risks of the green revolution as it is normally evaluated (in the policy time-frame) have been mentioned or discussed. Only a few additional points need to be made to round out the picture. One relates to the attempt of the United Nations family to extend planning and decision making to the full decade length that was somewhat arbitrarily set as the outer limit to the policy time-frame. The First Development Decade (launched in 1960 at the suggestion of President Kennedy) was characterized by a general international climate where conventional Western views on development tended to predominate, although there was an increasingly organized attempt on the part of the poor countries to try to obtain a larger share of the pie. The Second Development Decade was launched with less clarity as to the order of priorities. This was in part because a much larger laundry list of political, social, legal, scientific, and technological goals had been added to the economic ones carried over from the First Development Decade. It was also in part because a number of new issues were on the horizon—embodied in the form of conferences scheduled on the global environment and the law of the seas—which, many were sure, would transform the Second Development Decade in mid-stride. The addition of conferences on population and world food problems, and especially the Special Session of the General Assembly held in the spring of 1974 (to consider general resource and economic questions), all indicated that this transformation was in fact occurring. It would appear that this general state of flux regarding international priorities will continue (and will probably become more confused) until there is a fundamental reassessment of longer-term trends and a better outlining and awareness of the basically new context in which we find ourselves.

The discussion of international priorities as they relate to agriculture has tended to become somewhat more precise during this same period. This has been a result of the growing conviction that we are faced globally with the specter of mass starvation and that the top priorities relate to increased production and better use and distribution of what is produced. At the beginning of the First Development Decade, agriculture was seen as a subsidiary part of the overall process of industrialization. By the mid-1960s there was a "famine scare," with predictions of mass starvation by various dates, 1975 in one case and 1984 in another.[55] This led to a flurry of national and international activity in regard to agriculture. In the United States, the Johnson administration successfully pushed for a more broadly conceived P.L. 480, and the president's Science Advisory Committee was asked to prepare what turned out to be a three-volume study, *The World Food Problem*. The secretary-general of the Organization for Economic Cooperation and Development (OECD) prepared a report on "the Food Problem of the Developing Countries," and in 1968 the OECD released a report on *Aid to Agriculture in Developing Countries*. It was at the time that the World Bank began reconsidering its agricultural programs and lending priorities. The FAO began work on its massive *Indicative World Plan for Agricultural Development* (since renamed the *Perspective Study of World Agricultural Development*—evidently because some governments were frightened by the mention of a "world plan").

[55]The 1984 date appeared in a 1966 report of the USDA Economic Research Service. The 1975 date was given by William and Paul Paddock in their book *Famine 1975* (Boston: Little, Brown, 1967). See Brown (*Seeds of Change*), pp. x–xi for details.

By the late 1960s there was a revived optimism that the date of a Malthusian convergence of population and food curves could be postponed. The monsoon in India, which had failed in the mid-1960s, returned, and excellent growing weather prevailed. The HYV package was introduced on a massive scale into those irrigated regions that could adopt it quickly, and the results were impressive. After two lean years, the market for grains was excellent, and that, combined with the various subsidies and tax write-offs that had been used as incentives to encourage farmers to adopt the HYV package, provided very high profits. The role projected for agriculture at the start of the Second Development Decade was therefore one that tended to stress the progressive introduction of more "modern" agriculture. Its role in the development process was seen to be much greater than in the First Development Decade, and at least some attention was given to a few of the social and nutritional "side effects" of the green revolution. There was concern, both private and public, regarding what were expected to be the "second-generation" problems of the HYV package, such as how to encourage the developing countries to begin to diversify their production, how to improve local and regional marketing systems, how to reduce the many agricultural trade barriers of the rich countries so as to enable the poor countries to export their expected surpluses and thereby earn sufficient foreign exchange to provide for the necessary imports for further development, etc.[56] Since the beginning of the Second Development Decade, however, the situation has reverted to one of deep concern and, to some degree, confusion. The faltering of the monsoon in Asia has reduced production; the drought in the Sahel region of Africa has led to serious starvation; the global scarcity and resulting high price of grain has led the rich countries to sell their grain rather than make it available for food aid programs; the energy crisis has led both to a rapid increase in the price of fertilizers and to severe shortages in the developing countries—particularly in Asia.[57] A World Food Conference was hastily called in order to try to sort out these and related problems.[58] There is, however, still an underlying uncertainty

[56]For a full discussion, see Brown, the FAO *Provisional Indicative World Food Plan,* 2 vols. (Rome, 1970), and the Reports of the 14th–17th Sessions of the Conference of FAO.

[57]See footnote 35.

[58]An *ad hoc* World Food Conference was chosen rather than a special session of the FAO for several reasons. First, important nonmembers of FAO, especially the USSR, could be included (China was in the process of reassuming its seat in FAO). Second, such an

regarding such longer-term questions as whether the monsoon patterns in central Africa and south Asia are shifting because of a change in climate, what the effect of higher energy prices will be upon agriculture (in both the developed and the developing worlds), whether the Malthusian crunch is once again imminent, and so on. As will be argued below, these and other important longer-term questions can be comprehended only by using longer-term time-frames and the various units of analysis appropriate to them.

The Developmental Time-Frame

The food–population equation is perhaps the only developmental time-frame problem that has received serious and continued study. While there are a number of evolutionary dimensions involved—and most population graphs depict population growth over an evolutionary scale—there is general agreement that it is in the coming century (or sooner) that major adjustments or changes will occur. It is unfortunate that the general discussion of this problem has been so ecologically uninformed. On the population side of the equation, the anthropocentric assumptions of Western social scientists become very clear when you contrast estimates of food need based on human population figures with those that show the net biomass demand of countries calculated by including human and animal demand.[59] Another serious weakness in the discussion of the Malthusian equa-

---

ad hoc conference could bring a much wider range of viewpoints and expertise to bear on the various problems and could bypass some of the rigidities and bureaucratic narrowness that often characterize the activities of the specialized agencies.

[59]See Georg Borgstrom, *The Hungry Planet* (New York: Collier Books, 1967), pp. 1–23. Also, while there is a great deal of discussion of the "demographic transition" brought about by the Industrial Revolution, few demographers have examined natural history to see what sort of other occurrences may have influenced European population figures. For example, Colin Clark, in his book *Population Growth and Land Use* (New York: St. Martin's, 1967), points out that the single species of flea that carries the plague lives only on the black rat but not on the Norwegian gray rat: "In the eighteenth century, perhaps because it found the climate getting colder (which geographers tell us was the case), or perhaps by accident, it [the gray rat] began securing passage on ships, and finding its way to England and other Western European countries. It was a larger and fiercer animal than the black rat, whom it recognized and attacked as a competitor for food and shelter. In the fight between the two species, the black rat in Western Europe was quickly extirpated. It was this strictly exogenous zoological event which freed Western Europe from the plague and, in the long run, probably did more to determine the rise in Europe's population than any other single event" (p. 51).

tion is that most analyses are based upon either national or global averages. It is not the average person who succumbs to starvation or to a malnutrition-induced disease—it is a person who is at the low end of the distributional ladder. Without making the distributional question an integral part of one's analysis, both statistically and logistically, any resulting prescription will be seriously misleading. A final weakness is that few historical studies have attempted to examine the interactions between food and population growth. While some debate has developed as to whether past innovations in food production occurred at times of great population pressure or at times that were relatively free from it,[60] it is not clear whether a resolution of the question would have much bearing upon the situation we face in the coming century. More recently, computer models have been used in an attempt to include a wider range of variables influencing food and population growth rates. The most famous of these, *The Limits of Growth,* includes five basic elements—population, food production, industrialization, pollution, and consumption of natural resources—in its rather complex global model.[61] Although the model represents a serious and significant attempt to include more factors and interrelationships than previous studies, it suffers from the weaknesses mentioned above regarding the exclusion of animals in calculating net biomass demand and, more seriously, from the global averaging of what in many cases appear to be noncomparable data (the global averaging problems being the more serious ones).

Beyond the food–population equation, there are a number of other related environmental questions that need to be systematically studied. A basic one concerns the very high maintenance costs that can be expected as man progressively intervenes in complex ecosystems where the various flora and fauna have a long history of coadaptation

[60]The view that innovation occurred under population pressure is developed by Ester Boserup in her *Conditions of Agricultural Growth: The Economics of Agrarian Change under Population Pressures* (Chicago: Aldine, 1965). The view that in England the "agricultural revolution" took place during a time of relative population stability is presented in Eric L. Jones and Stuart J. Woolf, eds., *Agrarian Change and Economic Development* (London: Methuen, 1969). For a thoughtful overview, including other aspects of the food–population equation, see Joseph Hutchinson, ed., *Population and Food Supply* (Cambridge: Harvard University Press, 1969).

[61]Donella H. Meadows, Dennis L. Meadows, Jørgen Randers, and William Behrens III, *The Limits to Growth* (New York: Universe Books, 1972).

(see quote in Chapter 2, pp. 25–26). An already discernible cost relates to the various "seed banks" that have been established to try to preserve some of the irreplaceable germ plasms that are threatened with extinction as monocultures replace the mixture of genetically diverse species planted in traditional agricultural systems. Seed banks are clearly a second-best approach:

> Seed banks are in theory a reasonable way of halting genetic erosion, but in practice they tend to be underfunded, inadequate, and vulnerable to accidents or carelessness. A major Peruvian collection of corn germ plasm, one of the largest in South America, was lost when three refrigeration compressors failed. Even at CIMMYT, some of the early and irreplaceable corn collections from the 1940's were found to have been lost when the seed bank was reorganized in the mid-1960's.[62]

In addition to the risk of irreparable loss, many species cannot be stored easily in seed banks; even those successfully banked are removed from normal selective pressures, something of importance over the years.[63] Perhaps the greatest risk lies in the inherent instability of such biologically simplified systems as monoculture agriculture. The tragic costs of the Irish becoming dependent upon the potato were described in Chapter 2. That such risks still continue even in the United States can be seen in the 1954 loss of 75% of the durum wheat crop to rust and in the 1970 loss of some 15% of the maize crop to a leaf blight.[64]

Beyond the attempts of plant breeders to maintain a varied stock of genetic breeding materials and to keep ahead of the new plant diseases occurring through natural selection, there are the attempts of chemists to protect our highly simplified agricultural systems through pesticides. The longer-term maintenance costs here must include not only the capital costs of the pesticide industry but the environmental

[62]"Green Revolution (II): Problems of Adapting a Western Technology," *Science* 186(27 December 1974):1187.

[63]For a discussion of this question and a suggestion that the best approach would be to set aside some 100 sites around the world to preserve plant populations *in situ*, see H. Garrison Wilkes and Susan Wilkes, "The Green Revolution," *Environment* 14 (October 1972):32–39.

[64]"The epidemic was caused by a new race of the fungus which is particularly damaging to corn carrying Texas male sterile cytoplasm. In 1970 some 90% of hybrid corn grown in the United States carried this cytoplasm since it saved the cost and the inconvenience of hand-detasselling for seed production." Peter R. Day, "Genetic Vulnerability of Major Crops," *Plant Genetic Resources Newsletter*, No. 27 (February 1972), p. 2. This is a clear example of how commercial and administrative feasibility dominates in the world of agribusiness.

costs. The latter are extremely difficult to calculate because of a basic ignorance of all of their biological and ecological effects, many of which are context-specific. In addition to all of the various vested interests involved in the debate over pesticides, there is also an underlying difference in perspective between the ecologist, who is trying to find a mixture of biological means and cultivation practices that will control a particular pest in a specific ecosystem, and the universal-technological view of the pesticide industry, which wants a minimum number of chemical formulas to control a maximum number of pests.[65]

Ever-increasing energy subsidies to agriculture cannot continue at past rates; in fact, "both energy and land resource limitations make it impossible to feed the present world population a U.S. diet (69 percent animal protein) that is based on U.S. technology."[66] These difficulties apply especially to one of the major sources of energy subsidy—fertilizer. In many irrigated areas, applications of fertilizer have reached such a point of diminishing returns that it would take 6.5 times as much fertilizer as now applied to double their production.[67] Beyond this, limited reserves of phosphorous (one of the three basic constituents of fertilizer) plus the fact that there is no process for synthesizing or reconstituting it, strongly suggest that it may become the limiting factor in HYV production:

> At present rates of population growth and phosphorus consumption, known phosphorus reserves will be used up in 60 years, by which time there will be 11,000 million people on this planet. Without commercial phosphate fertilizers, however, the planet will be able to support only 2,000 million people.[68]

The longer-term implications of expanding irrigation systems are also unclear. The history of water use in the developed countries has been one of using water more rapidly than its replacement by hydrological recharge rates. What this has led to in much of the western

[65]For a good summary of the ecological costs of pesticides, see Raymond F. Dasmann, John P. Milton, and Peter H. Freeman, *Ecological Principles for Economic Development* (New York: Wiley, 1973), pp. 149–166. One example is the creation of new pests: "In Kenya, an outbreak of the giant coffee looper in coffee plantations was traced to the lethal effect insecticides had on its parasite. Neither the giant looper nor its parasite were target species" (p. 151).

[66]Pimentel, p. 760.

[67]*Man in the Living Environment*, p. 17.

[68]Ibid., pp. 25–26. In addition, there are serious water contamination and eutrophication problems associated with high fertilizer use.

United States is declining water tables, leaching, and salinization, plus the construction of huge canal systems to "import" water from wetter regions.[69] Water tables in India and Pakistan appear to have declined, particularly in those regions where the HYVs are grown. Beyond this, there are some indications that large dams and increases in irrigation (as well as the other land-use changes associated with the HYVs) may have an impact upon climate, rainfall, and monsoon patterns. The normal longer-term changes in climate referred to at the beginning of Chapter 2 make it particularly difficult to estimate the influence of any one factor. In any case, several climatologists have suggested that we are moving out of the particularly favorable climatic conditions that agriculture has enjoyed for the last 30 years. They foresee a period when weather in the northern temperate zone will become colder and when there will be significant shifts in the latitudinal flow of the monsoons.[70] While several recent international programs such as the Global Atmospheric Research Program (GARP) and various sampling and satellite programs provide a data base for mapping changes as they take place, there is much less in the way of a historical data base for comparisons. One of the reasons for this is the general low priority given to studies of natural history and the multidisciplinary requirements for getting a fairly complete overview.[71]

Mention of the impact of land-use patterns upon climate and vice versa leads to a brief consideration of other aspects of land use. There are many questions and problems related to soil. The list includes deforestation, soil erosion, the generation of dust through agriculture, overcompaction from heavy machinery, problems of how to provide

[69]In Santa Clara County, California, declining water tables have caused the land to subside some 12 feet in the last 50 years. Ironically, a similarly caused subsidance seriously interrupted the gradient of the Delta-Mendota Canal in Central California (heavy irrigation being introduced after the canal was built). Committee on Geological Sciences, National Academy of Sciences, *The Earth and Human Affairs* (San Francisco: Canfield Press, 1972), pp. 70–71.

[70]See Reid A. Bryson, "A Perspective on Climatic Change," *Science* 184 (17 May 1974):753–760, and Hubert H. Lamb, *The Changing Climate* (London: Methuen, 1972). In the United States, a 20-year drought cycle has been noted since the Civil War. If the 1970s repeat the pattern of the droughts of the 1930s and 1950s, this may compound any longer-term shifts.

[71]Material must be synthesized from such varied sources as studies of tree rings, studies of fossilized pollen, an examination of agricultural histories, and the location and comparison of sketches and paintings of glaciers that were made at different times.

the developing countries with soil surveys, the difficulties in analyzing nutrient cycles and rates of uptake, and a whole range of political, legal, and economic problems that go back to the central position of land control in the distribution of power, influence, and economic goods.[72] Finally, there are questions about differing cultural perceptions of nature, land, and agriculture. Some appreciation of the importance of these can be gained by noting the very different ways in which the ideas of economists and ecologists challenge certain "traditional" views. The ecologists

> are suggesting that we, the rich Western countries, must change our basic attitudes towards nature, our modes of production, many of our institutions, and many of our comfortable habits. The challenges which these "subversive scientists" present to our traditional ways are strikingly similar to the challenges which conventional economists have presented to the non-Western world in urging them to "develop." It is only when we think about the degree and difficulty of change that is being asked for by the ecologists (as well as the social and institutional resistance to such changes) that we can begin to appreciate the ways in which conventional development processes challenge the poorer countries of the world.[73]

In spite of a number of serious studies by environmentalists and historians on different cultural attitudes toward nature and land,[74] little of this work seems to have penetrated the thinking of most social scientists. This is partially the result of their meager interest in agricultural matters (an attitude that mirrors the low value placed on agriculture by most industrial societies) and partially the result of the short time-frames and universalistic assumptions they use.

The various problems brought up in this section are simply illustrative of a much wider range of issues that emerge in looking at things in terms of a developmental time-frame. No systematic attempt is

[72]For a comparative historical review of the complex interactions within several countries and the wide variation in patterns between those countries, see Barrington Moore, Jr., *Social Origins of Dictatorship and Democracy: Lord and Peasant in the Making of the Modern World* (Boston: Beacon Press, 1966).

[73]Kenneth A. Dahlberg, "Ecological Effects of Current Development Processes in Less Developed Countries," in *Human Ecology and World Development*, edited by Anthony Vann and Paul Rogers (New York and London: Plenum Press, 1974), p. 71.

[74]For example, see Clarence J. Glacken, "Changing Ideas of the Habitable World," in *Man's Role in Changing the Face of the Earth*, edited by William L. Thomas, Jr. (Chicago: University of Chicago Press, 1956), pp. 70–92, and Robert E. Frykenberg, ed., *Land Control and Social Structure in India* (Madison: University of Wisconsin Press, 1969).

made here to develop units of analysis, concepts, and propositions that would help to sort out and rank the relative importance of the different questions. Instead, I am trying to demonstrate that there are a whole range of costs and benefits that, although critical to any overall evaluation of the green revolution, have for one reason or another been neglected or left out. It would appear that one of the major reasons for this is the particular way in which most Westerners understand "development." Another reason is the type of standards that are used to assess that development. The increasing use of energy measures to reevaluate economic problems suggests that the dominance of purely economic measures and modes of thought may soon be challenged.[75] The very fact that we can now use energy measures to analyze a good many happenings in the industrial world reflects a transformation in the use of energy toward greater degrees of interchangeability between energy sources:

> It would have made no sense to measure societies in terms of energy flow in the 18th century when economics began. As recently as 1940, four-fifths of the world's population were still on farms and in small villages—most of them engaged in subsistence farming. In the 18th century commercial farms were known. The very existence of cities depends upon a kind of commercial farming that produces a surplus of food for church or other group that controls such farms. But that was also the reason why the measure of energy would have made no sense at the onset of the 19th century. Commercial farms of that era were controlled by church, traditional aristocrats or other groups for their purposes. And if they might, at times, participate in the economic marketplace, they did not—and mostly could not—exchange energy. Only after some nations shifted large portions of the population to manufacturing, specialized tasks, mechanized food production, and shifted the prime sources of energy to move society to fuels that were transportable and useable for a wide variety of alternative activities could energy flow be a measure of societies' activities. Today it is only in one-fifth of the world where these conditions are far advanced.[76]

It would thus appear that we are entering a new period (or "development play") where major changes in thinking and habits will be required.

[75]For recent examples, see Bruce M. Hannon, "An Energy Standard of Value," *The Annals* 410(November 1973):139–153; John S. Steinhart and Carol E. Steinhart, "Energy Use in the U.S. Food System," *Science* 184(19 April 1974):307–316; and Howard J. Odum, *Environment, Power and Society* (New York: Wiley, 1971).

[76]Full text from prepublication version of the article by the Steinharts, p. 307.

The Evolutionary Time-Frame

One fairly pervasive habit that needs to be guarded against in discussing the evolutionary time-frame is the tendency to project into the future the particular configuration of forces (or the conventional perceptions thereof) that happen to exist at the moment. The same problem as it applies to the past was discussed in Chapter 2, where I attempted to show that there is considerable variation in the information forwarded over time in the genetic, ecological, and cultural channels (see Figure 2). Equally, the basic elements involved in agriculture—climate, soil, plants, animals, and men—were all shown to have exhibited significant variation when examined in evolutionary perspective. There has been not only variation but evolution. The role of man as a geologic agent has been mentioned; in Chapter 5 this is examined in more detail, to show the complex feedback effects of man's agricultural and industrial activities upon the hydrologic cycle. What becomes clear is that large-scale units of analysis like the hydrologic cycle, global deforestation rates, and amounts of atmospheric dust, as well as man's progressive impact upon them, must be incorporated into our thinking. This goes well beyond traditional approaches to natural history and most current attempts at developing theories of cultural evolution.[77] It also runs counter to the dominant tendency of increasing disciplinary specialization. Unfortunately, disciplinary specialists rarely realize the risks of ignoring parametric variation (until too late) and appear not to be fully aware that such large-scale "maps" of the evolutionary interactions between man and the biosphere are necessary to give them an idea of where they are and what the meaning of their detailed work is.

It would appear, then, that the appropriate approach to take in

[77]For useful overviews, see Peter A. Corning, "The Biological Bases of Behavior and Some Implications for Political Science," *World Politics* 23(April 1971):321–370, and Claude S. Phillips, Jr., "The Revival of Cultural Evolution in Social Science Theory," *Journal of Developing Areas* 5(April 1971):337–370. As Corning points out (p. 354), "While certain evolutionary trends have been postulated at one time or another (such as decrease in entropy, maximization of metabolism, minimization of effort, increase in homeostasis, a growth in complexity, an increase in cooperation, and progress in feeling, knowing, willing, and understanding), there is no agreement on this point among biologists."

looking toward the future is to try to remain sensitive to longer-term variations, to be aware of the basic parameters of ecosystem change, and to keep as many evolutionary options open as possible. Mention has already been made of the risks of reducing the genetic diversity of man's major food crops. The basic ecosystemic parameter that is involved here has to do with the increasing instability of ecosystems as they are simplified (i.e., as the number of different species is reduced).[78] Beyond the risks of plant diseases, which have been discussed as a risk analyzable in terms of the developmental time-frame, there are other kinds of risks associated with the ecosystem simplification produced by the green revolution. First, there is the clear implication that as the human population continues to increase (in part encouraged by increased agricultural production), many more plant and animal species will become extinct.[79] The suggestion here is that by simplifying his agricultural systems to try to increase production in the short run, man ends up with an unstable agricultural system and also seriously simplifies the global ecosystem. While one can give only rough probabilities for any of the disastrous consequences that may result, this does not make them any less real or less important.[80]

In talking about keeping evolutionary options open, I am talking not only about maintaining the diversity of our life-supporting environment, but also about how to maintain human and social adaptability. The basic institutional question involved here is whether such adaptability is compatible with the increasingly more standardized and global pattern of modern industrial society—something that

[78]See Marston Bates, "The Human Ecosystem," in the National Academy of Sciences study on *Resources and Man* (San Francisco: Freeman, 1969), pp. 27–28.

[79]For a full discussion, see Jean Dorst, *Before Nature Dies* (Baltimore: Penguin Books, 1970), especially Chapter 2.

[80]The following are some consequences that have received discussion: that DDT levels (along with other pollutants) in the oceans might increase to the point of seriously reducing the oxygen-producing *phytoplankton* or seriously interfere with reproduction of major fish species; that eutrophication of the oceans might seriously change all oceanic life cycles; that increased farming on marginal lands risks not only soil erosion but the generation of sufficient atmospheric dust to modify the climate. Of course, there are many additional long-term dangers, such as heavy-metal poisoning and contamination from radioactive wastes, which are a result of high levels of industrial activity.

Boulding has called the "superculture."[81] While most people hold to the idea that industrial society is more complex than previous ones, it is not clear that this is the case, especially if one analyzes it in terms of either energy use or agriculture. The shorter-term risks of industrial countries' becoming too dependent upon one cheap but nonrenewable energy source have been strongly dramatized as a result of the OPEC oil boycott. The risks of converting the major portion of our agricultural production to very-high-energy modes that are dependent upon nonrenewable resources appear even greater.[82] An appropriate evolutionary approach—trying to increase sustainable production while maintaining or increasing diversity—will be examined in Chapters 5 and 6. Such an approach goes against many of the intellectual and institutional habits of modern industrial society. This returns us to the basic problem of adaptability and the question of whether we are going to let our current habits and vested interests carry us down paths that are evolutionary dead ends, or whether we are going to try to respond to those who are calling for the kinds of change that will help us keep our evolutionary options open. It is only those who have thought with some care about evolutionary questions who will be able to suggest the paths toward survival.

[81]Kenneth E. Boulding, "The Interplay of Technology and Values: The Emerging Superculture," in *Values and the Future,* edited by Kurt Baier and Nicholas Rescher (New York: Free Press, 1969), pp. 336–350. He describes it as follows: "The superculture is the culture of airports, throughways, skyscrapers, hybrid corn and artificial fertilizers, birth control, and universities. It is worldwide in scope; in a very real sense all airports are the same airport, all universities the same university. It even has a world language, technical English, and a common ideology, science."

[82]One risk is that of a production–population bubble, which is not sustainable and collapses (cf. the discussion of phosphorus reserves above, p. 82). On the other hand, "Studies of animal populations suggest that environmental factors other than simple limitation of material resources may act in unexpected ways to limit populations before theoretical maxima are reached. To consider whether the earth might support three more doublings of the human population is probably to consider a purely hypothetical situation. It seems more likely that future crowding, the necessary social and governmental restrictions that accompany dense settlement, and certain kinds of boredom resulting from isolation from nature in an immense, uniform, secular society may prove so depressing to the human spirit or so destructive of coherent social organization that no such population size will ever be reached" (*Resources and Man,* pp. 8–9).

SUMMARY

We should perhaps return to the question posed at the beginning of the chapter: Are we or are we not engaging in a "foolish bargain," such as Jack's? It all depends upon how you look at it—that is, it depends upon whether you evaluate the green revolution strictly in terms of a policy time-frame or whether you also evaluate it in terms of the developmental and evolutionary time-frames. While all agree that there have been a number of positive results, even staunch supporters of the green revolution (who tend to look only at the policy aspects) have discerned a number of serious "side effects," such as economic and regional polarization, increasing rural unemployment, and little positive impact upon rural poverty. When longer-term questions or criticisms are raised, one of the basic moral dilemmas of the world today is posed in response: Can you condemn millions to starvation by not making every effort to increase agricultural production, especially if there is any chance that some solution to the population problem can be found in the coming decades? To answer this directly would require 100% confidence that no solution to the population problem can be found and that consequently you must decide whether it is better to face the crunch now or later, when hundreds of millions more would face starvation.[83] What is most important to recognize is that there are options and approaches other than those posed in terms of either more production through the green revolution or "lifeboat ethics." Many such alternatives are discussed in Chapters 5 and 6. Briefly, the dilemma as posed illustrates the lack of imagination and the power of habit that are embodied in the green revolution: we simply find it the easiest approach because it represents, with minor adaptations, an exporting of our technologies and views and places the burden for increasing production primarily upon the poor countries. Alternative options—which might include, among other things, a changing of our tariff and quota policies, a changing of our basic approach to foreign aid, a shift from our meat-laden diet, a shift in the distribution of

[83]Those discussing the possibility of cutting off aid to those countries that are not "savable" include William Paddock and Paul Paddock, *Famine 1975!* (Boston: Little, Brown, 1967) (with their famous "triage" approach), and Garrett Hardin, with his various formulations of the "lifeboat ethic."

wealth, a rethinking of our ideas about technology, and a decentralization of a number of our institutions—all would require *us* to make major changes. It is not surprising that the green revolution approach, which requires *them* to change, is the unconsciously preferred option.

It is when the green revolution is looked at in terms of the developmental and evolutionary time-frames that you must conclude that it is a poor bargain. Given the risks of plant disease, pests, exhaustion of critical resources (especially fuels and fertilizers), and soil erosion, not to mention its associated social, political, and economic dislocations, it is doubtful that modern industrial agriculture is sustainable over the next century. The developmental risks may be compounded by evolutionary shifts, particularly that of climate. There is also the evolutionary risk of so simplifying our agricultural and global ecosystems that there is some sort of serious biological collapse—which will have direct consequences upon whatever number of billions of humans are then alive. The overall conclusion need not be one of doom but simply that we must trade in our current models (both intellectual and agricultural) for smaller, better, and more ecologically sound ones.

4

The Momentum of Structures, Institutions, and Current Policies

When, after many efforts, a legislator succeeds in exercising an indirect influence upon the destiny of nations, his genius is lauded by mankind, while, in point of fact, the geographical position of the country, which he is unable to change, a social condition which arose without his co-operation, customs and opinions which he cannot trace to their source, and an origin with which he is unacquainted exercise so irresistible an influence over the courses of society that he is himself borne away by the current after an ineffectual resistance. Like the navigator, he may direct the vessel which bears him, but he can neither change its structure, nor raise the winds, nor lull the waters that swell beneath him.

Alexis de Tocqueville, *Democracy in America*

The analysis so far has focused on the global and long-term development of agriculture, especially modern industrial agriculture. The logic of contextual analysis also requires an examination of how structures and institutions at the national and international level shape, channel, or distort policy making. The influence and momentum of these structures and institutions needs to be understood if we are to appreciate how we got where we are as well as to make a realistic assessment of how alternative policies might be pursued. The emphasis in what follows is on the developing countries—the recipients of the green revolution—and upon the types of influences that shape their decisions.

THE DECISION-MAKING CONTEXT OF NATIONAL ELITES

The long-term or evolutionary forces with which the decision maker finds himself confronted can be described in general terms, but it is difficult, if not impossible, to trace their specific influence on particular policy questions. As argued in the preceding chapters and as traced out in Arnold Toynbee's work on civilizations, it is most important to know the interactions and relationships between nature, man's modifications of it, and the basic social structures of groups of civilizations.[1] However, at this point, we need to examine the influences, processes, structures, and institutions that operate at the level of the developmental time-frame. Although neglected in much of the literature on economic development, the most prominent of these—cultural influences, political-administrative influences, and external Western influences on elites—are of basic importance and can be traced with some precision.

Cultural Influences

Although there is no clear agreement on definitions, it would appear that the concept of culture fits the developmental time-frame more appropriately than the evolutionary one.[2] This is because discussion of specific cultures or changes in the culture of individual coun-

[1]See Arnold J. Toynbee, *A Study of History* (New York: Oxford University Press, 1948–1961). Care must be used to avoid some of the Western biases associated with the term *civilization*.

[2]Analyzing various concepts in terms of the different time-frames would be most useful in clarifying ambiguities and sorting out competing definitions. On the one hand, there would be the question of what the most appropriate concepts are for a given time-frame. On the other, there is the question of whether a particular concept silently shifts from one time-frame to another as it is applied to specific situations. The various concepts of culture are particularly subject to this silent shell game, and the contrasting definitions of anthropologists like Leslie White, Julian Steward, Marshall Sahlins, Elman Service, and Marvin Harris would all benefit from attempts to relate them to specific time-frames and units of analysis.

tries generally involves analyzing events that take place over the span of a century or less.

In discussing cultural influences, one must be careful to avoid making an assumption fairly common in the literature on economic development—that the critical question is how to overcome cultural inertia:

> The assumption of inertia, that cultural and social continuity do not require explanation, obliterates the fact that both have to be recreated anew in each generation, often with great pain and suffering. To maintain and transmit a value system, human beings are punched, bullied, sent to jail, thrown into concentration camps, cajoled, bribed, made into heros, encouraged to read newspapers, stood up against a wall and shot, and sometimes even taught sociology. To speak of cultural inertia is to overlook the concrete interests and privileges that are served by indoctrination, education, and the entire complicated process of transmitting culture from one generation to the next.[3]

Another point to be kept in mind is that most of our information and understandings of other cultures have been filtered and structured according to Western conceptions and categories:

> Non-Western societies have not brought forth equivalents either to our general structured universe of knowledge, or to the particular academic disciplines and methodologies of which it is composed, just as they have not produced "Occidentalists" or other foreign area specialists to match the legions of Orientalists and Africanists that have for centuries been at work in our civilizations.[4]

Even if one is cautious about hidden Western biases and the tendency to focus on change more than on the problems of socialization, there are formidable problems in trying to trace out the multicultural interactions and influences at work in the non-Western world. By looking at the most recent "developmental plays," we can sort out some of the major influences upon succeeding generations of elites. While there is not space here to present case histories of particular countries or regions, the kinds of issues that need to be dealt with can be raised and some appropriate modes of analysis suggested.[5]

[3]Barrington Moore, Jr., *Social Origins of Dictatorship and Democracy: Lord and Peasant in the Making of the Modern World* (Boston: Beacon Press, 1966), p. 486.

[4]Adda B. Bozeman, *The Future of Law in a Multicultural World* (Princeton, N.J.: Princeton University Press, 1971), pp. ix–x.

[5]The magnitude of effort involved in any case study synthesizing the issues to be raised here can be seen in Gunnar Myrdal's monumental regional study *Asian Drama: An Inquiry into the Poverty of Nations* (New York: Twentieth Century Fund, 1968).

The stage settings for recent developmental plays in the non-Western world were largely made up of elaborate structures imported from the colonial countries—structures that thereafter strongly influenced succeeding generations of native elites. The general thrust of these imported structures was to establish some fairly standard and comprehensible Western pattern that it was hoped would eventually replace the great complexity and diversity of cultural patterns found in most colonies. These structures included both infrastructural projects like roads and canals, which had an impact on the natural environment, and institutional reforms that gradually changed the broad social environment. There were, of course, interactions between them as well.

One example of the environmental impact of infrastructural projects was given in Chapter 2 in regard to railroads and canals built in colonial India. More generally, infrastructural projects tended to structure the transport, shipping, and market activities of the colonies according to the commercial purposes of the metropolitan countries. What this normally meant was that the road, railroad, communication, and market networks were funneled into the one or two major exporting ports of the colony. There were few infrastructural links between colonies, and the paucity of such links even today has hampered the development of regional trade, particularly in Africa. The arbitrary and often fortuitous way in which many colonial boundaries were drawn also has had profound influences upon environmental management and conservation in the former colonies. The division of river basins, watersheds, cultural groupings, and preexisting market areas has meant that their governance and management often become issues between neighboring states.[6]

The institutional changes and reforms molded in colonial times have had great durability and continue to shape the basic context of decision making in most of the non-Western states. Traditional legal systems were replaced or subsumed in the newly established Western systems. Financial and marketing systems were modified or set up to meet the needs of the metropolitan powers. Educational systems were imported, either by missionaries or later by civil servants, first to

[6]For a discussion of different cultural perceptions of boundaries and their meaning, see Robert Montagne, "The Nation-State System in Modern Africa and Asia," in *Comparative World Politics,* edited by Joel Larus (Belmont, CA: Wadsworth, 1964), pp. 58–71.

"civilize," then to help train local administrators to assist in the running of the colony.[7] Governmental and administrative systems varied, but all were structured to facilitate the achievement of the major commercial and strategic goals of the metropolitan power. There have been any number of typologies proposed in the literature on economic development to try to describe the relative degree of impact that Western societies have had upon the non-Western world. Many of these propose a continuum of "progress" ranging from traditional societies (little Western impact) to transitional societies (major Western inroads) to developed societies. The built-in assumption that all other societies should be "progressing" toward the Western industrial model has often been noted. While there are important pressures toward global technological standardization and toward some sort of superculture (at least among a small elite), it is clear that for a long time "the world will continue to be multicultural under the surface of unifying technological and rhetorical arrangements"[8] and that any major non-Western cultural renaissance could seriously undermine many current international institutions and practices.

An additional and equally serious deficiency in most typologies of economic development is that they generally fail to explain the interacting dynamics of change and cultural continuity. If the history of colonialism is examined in terms of "developmental plays," it soon becomes clear that the key actors in each play are the successive generations and that the dynamics of change and continuity can be

[7]This was generally the case for the French, British, and Dutch colonies. The Belgians tended to have a broader-based elementary education, but fewer students at the high school or college level. The United States in the Philippines sought to establish a replica of its own mass education system. Myrdal's assessment (Vol. 3, p. 1647) is as follows: "The other colonial powers rather neglected popular education, while the secondary schools and colleges they established usually emphasize literary and academic training. Their objective was, not to change the people's basic attitudes and help prepare them for economic development, but to train docile clerks, minor officials of all sorts, and, particularly and increasingly in the British colonies, higher administrative functionaries. Insofar as there was a zeal to spread Western civilization—and on the part of the British and the French, there undoubtedly was—this was focussed on "higher" culture and on the upper strata. The policy of giving a literary and academic rather than a vocational character to most of the new schools was also in line with the colonial power's disinterest in encouraging indigenous manufacturing industry in their possessions."

[8]Bozeman, p. 163.

greatly clarified by looking at the factors and context shaping each generation as well as at the interactions between generations.

The concept of generations has received relatively little attention in modern social science. Some statistical comparisons are made in sociology, and there is occasional reference in political science to the difference between the "first" and "second" generation leaders in the newly independent countries of Asia and Africa. Few systematic attempts have been made to examine the utility of using generations as a unit of analysis or to trace the methodological difficulties involved. One of the latter, a result of the basically different cultural contexts that different societies (and thus generations) find themselves in, is pointed out by Margaret Mead. In a series of lectures discussing the generation gap, she distinguishes three different cultural contexts within which children grow up: "*postfigurative,* in which children learn primarily from their forebearers, *cofigurative,* in which both children and adults learn from their peers, and *prefigurative,* in which adults also learn from their children. . . .[9] She is thus emphasizing both the importance of education in the socialization of generations and the differences to be found between cultures regarding such education.

A more systematic approach is found in the work of the Spanish social philosopher Julian Marias, who builds upon the work of Ortega y Gasset. He stresses that the concept of generations must include biological, social, and historical components. He sees a pattern of four 15-year generations, each with a specific sociopolitical function: (1) the "survivors," or elder statesmen; (2) those in power (who reflect the prevailing world style); (3) the "opposition," or rising generation; and (4) youth.[10] The social aspect is to be found not only in educational patterns but in the *vigencia* of a society or collectivity—those prevailing customs, laws, usages, traditions, and beliefs that permeate each generation as it grows up.[11] The historical components relate both to the ways in which generations are interlocked like "tiles on a roof" and to the ways in which generations and *vigencia* alike are influenced and

[9]Margaret Mead, *Culture and Commitment: A Study of the Generation Gap* (Garden City, N.Y.: Natural History Press, 1970), p. 1.

[10]Julial Marias, *Generations: A Historical Method* (University: University of Alabama Press, 1970), pp. 183–184. He recognizes that increasing longevity may create a fifth generation.

[11]Ibid., pp. 81–83.

changed by major historical events, such as wars, revolutions, and depressions.[12] Marias's approach is generally compatible with the type of contextual analysis proposed in Chapters 1 and 2. In addition, it stands as an example to social scientists of how units of analysis should be related to the time-frame chosen for or implicit in a work.

To actually apply the insights of Mead and Marias to the 19th and 20th centuries would require a major shift in historical analysis. Different societies would have to be examined separately both in terms of the interactions between generations over time and in terms of the major internal or external influences that modify the basic context of that society or have a major cultural impact upon one or more of the generations. While some broad Western influences upon the non-Western context have been mentioned here, it should be recognized that in a full historical analysis such influences would be seen to hit different societies at different times and with different force.

The same qualification applies to the following discussion of one of the major conduits of cultural influence in the 19th and 20th centuries: Western educational systems. The structure and content of these systems has had and will continue to have great influence upon the non-Western countries, particularly in regard to agriculture:

> In colonial times, the school system of the metropolitan country was copied, in diluted form, in each of the South Asian territories. It has been changed only slightly during the period of independence. Like all institutional structures, the system embodies strong vested interests. . . . The widespread disregard of adult education discussed in the preceding chapter is, for instance, a reflection of educational attitudes inherent in the structure of formal schooling inherited from colonial times, which everywhere but in the Philippines was designed mainly to create and preserve an elite society. To a greater extent than in any other field, foreign influences on the school system, even after independence, have come almost exclusively from the Western countries.[13]

In some cases, like India, the elite nature and formalistic methods of education blended well with traditional educational systems, which were also elitist and formalistic. However, the cultural content and focus were greatly transformed. As a result, each generation of elites has tended to emphasize, perhaps with a certain time lag, those themes of education and techniques of social control dominant in the

[12] Ibid., pp. 154–155, 187–188.

[13] Myrdal, p. 1697.

West. They have thus been able to increase their personal social mobility while at the same time gaining the necessary skills to pursue the broader national goals of greater political and economic freedom. In the 19th and early 20th centuries, law was generally the dominant theme, although since independence the emphasis has tended to shift to economics.[14] While there has been some utility in drawing upon these Western models, notably in the achievement of independence, the costs of Western educational systems have been very high in the agricultural field. The models imported, even in the 19th century, tended to reflect only a part—the most academic- and urban-oriented part—of the educational systems of countries already well along the path to industrialization. As a consequence, they did not then nor do they now fit the largely agricultural needs of the developing countries.

The academic and urban content of education and its generally elitist structure have also combined with traditional status attitudes and psychological pressures to produce elites that often have anti-manual-labor and antiagricultural attitudes. In regard to manual labor, Myrdal points out that:

> Throughout South Asia there is a traditional contempt for manual work, and the educated tend to regard their education as the badge that relieves them of any obligation to soil their hands. In India, the Brahman prescriptions against plowing and various other types of manual work have, of course, helped to strengthen this prejudice; but it is prevalent to some degree everywhere in the region.[15]

Urban elites, virtually across the political spectrum, have tended to see agriculture, and particularly traditional peasant agriculture, as an unpleasant reminder of the traditional past and of all the obstacles that they face in trying to "modernize" their countries. In the past few years, there has been some increased intellectual awareness of the need to devote more developmental resources to improving the agricultural sector, but the tendency has been to draw upon Western models of industrial agriculture. Until recently, little has been done, or even proposed, to try to reform (or revolutionize?) the educational

[14]In discussing the shift away from law, Bozeman, p. 163, points out that "few of the personalities now dominant or ascendant on local planes of politics have had, or are likely to have, the kind of intellectual connections with the legal thought world of the West that had stimulated many of their predecessors to think of 'national self-determination' in the context of law and constitutionalism."

[15]Myrdal, p. 1646.

systems of developing countries to fit them to their largely rural and agricultural needs.[16]

A recent study commissioned by UNICEF gives some indication of the major changes required in educational priorities and practices to be able to meet these needs.[17] Given the inflexibility of the formal systems of education and the economic pressures from which they will increasingly suffer, the report argues that the best strategy is to encourage a variety of nonformal approaches to education. The groups that have been most neglected, and toward whom nonformal education should be directed, are pre-school-aged children, out-of-schoolers (adults as well as dropouts), and girls. While the need for flexibility is stressed, the report argues that there are six goals that should be included in any "minimum package":

1. Positive attitudes
2. Functional literacy and numeracy
3. A scientific outlook and an elementary understanding of the processes of nature
4. Functional knowledge and skills for raising a family and operating a household
5. Functional knowledge and skills for earning a living
6. Functional knowledge and skills for civic participation[18]

These elements, while aimed at neglected groups and designed to meet many of their needs, still include (through the rather visible hand of John Dewey) a number of Western assumptions about education as an agent of social change. However, the two most difficult steps required to realize such recommendations lie largely outside the capabilities of any aid program (bilateral or international). First, any major shift in governmental allocations toward nonformal education requires both the political power and the will to overcome existing

[16]A typical example of neglect can be found in the fact that the index to James Coleman's, ed., *Education and Political Development* (Princeton, N.J.: Princeton University Press, 1965), does not contain a listing for either agricultural or rural education (although some of the individual articles in the volume do contain some discussion of these matters).

[17]Philip H. Coombs, Roy C. Prosser, and Mazoor Ahmed, *New Paths to Learning for Rural Children and Youth* (New York: International Council for Educational Development, 1973).

[18]Coombs, pp. 14–15.

vested interests. Second, the task of designing genuinely functional programs is one requiring such local cultural and environmental knowledge, as well as ability to adapt programs to their variations, that few outsiders can be of much help (other than perhaps sensitizing program designers to some of the cultural and environmental dimensions that need to be included).

The dilemma is clear: the academic- and urban-oriented elites in the developing countries are unlikely to pursue educational reforms along the lines suggested by the UNICEF report with any great enthusiasm. The elites in the industrial countries, who might be able to influence the elites in the developing countries by means described below, are equally unlikely to recognize on a widespread basis the inappropriateness of many of the intellectual and technical models that they export. To do so would reduce their leverage and would also require a reassessment and devaluation of the prestige and perquisites of external experts.

Political-Administrative Influences

The variety of Western political and administrative influences upon the decision-making context of local elites can be delineated in much the same manner as the cultural influences. First, there are a number of longer-term influences that are observable in the different historical "development plays." The general impact of colonialism upon the environment, the infrastructure, and the institutions of many developing countries has been referred to. Among the political and administrative influences, one of the most pervasive longer-term influences has been the importation of political and administrative structures that assume a high degree of cultural homogeneity and general agreement about the central role of law in society, neither of which was present in most receiving colonies. Even upon independence, what one found was a host of states that either contained a number of major ethnic groupings or balkanized a major national group.[19] In either case, the problems of the new governing elites in trying to develop new

[19]For a detailed discussion of the difficulties created in Southeast Asia by discontinuities between political, legal, and cultural boundaries, see Clifford Geertz, "The Social–Cultural Context of Policy in Southeast Asia," in *Southeast Asia: Problems of United States Policy*, edited by William Henderson (Cambridge, Mass.: MIT Press, 1963), pp. 45–70.

loyalties to the state that would transcend the preexisting and usually more localized loyalties to religion, caste, village, tribe, clan, or family have been complicated by the inappropriate nature of these imported institutions as well as of the other environmental and infrastructural legacies of colonialism.

An assessment of the momentum of the institutional structures introduced by the colonial powers in the political and administrative fields is more complex than for educational institutions. While each generation shares similar vested interests and concerns about social mobility and status, there have been major divisions, shifts, and changes in the political and administrative arenas. The most fundamental of these—the change at independence in their dominant purposes and dominant personnel—raises the question of whether the Western character of these institutions will not gradually be replaced, leaving only a Western facade.[20] The gradualness of the change can be related to the differential rates at which "second-generation" leaders come to control not only the government but also the various bureaucracies. There also appear to be significant variations in many countries as to which ministries change, how much they change, and in which direction. In some areas, particularly in the technical services, there appears to have been little change between the pre- and postindependence periods either in the general operating routines or in the relative status of the different technical branches.[21] In other areas, particularly in foreign policy, great changes have been made and new institutions

[20]As Bozeman, pp. 164–165, phrases it: "Although law is of minimal importance in all non-Western societies, organization is everywhere of the essence. However, in this context too one notes a dual frame of reference; for under the surface of organizational structures that are analogous in outline, even if not in inception and function, to those found also in the West, a great variety of very different associational forms exist and are likely to remain viable in response to local needs. Thus it appears clear after an analysis of modern Asian and African states that 'the nation' is not a widely meaningful composite concept, that the territorial contours of the state—be it a mini-state or a quasi-imperial orbit—can be indeterminate and shifting without detracting from the identity of the political organism, and that the cause of the central government is not necessarily linked to principles denoting either internal stability or popular representation and respect."

[21]It has been pointed out that the Indian Canal Department, which was the senior technical service in the 19th century, continues to be so today. See Elizabeth Whitcombe, "The New Agricultural Strategy in Uttar Pradesh, India, 1968–70: Technical Problems," in *Technical Change in Asian Agriculture,* edited by Richard T. Shand (Canberra: Australian National University Press, 1973), pp. 183–201.

and routines have been created to enable these countries to respond more autonomously to the world. In some areas, particularly those dealing with economic and military matters, the influence of industrial countries is quite apparent and often encouraged.

The spatial distribution of administrative services, if not political structures, tends to follow the infrastructural patterns set up during the colonial period and elaborated upon since independence through various aid and development programs. The funnel-like character of most infrastructures has been mentioned. Increasingly, geographers have become interested in tracing the striking influence of such patterns upon the spatial distribution of economic development and modernization.[22] One detailed study traces from colonial times onward the ways in which modernization in Kenya has been correlated spatially with the major road, railroad, communications, and administrative networks; this work also suggests the caution that must be exercised in using development indices based upon national averages—for without a clear idea of their spatial distribution, they become misleading.[23] It would be interesting to know to what degree the spatial distribution of military units also conforms to these infrastructural patterns and whether strategic considerations modify them in any significant manner.

The combination of high levels of influence from the industrial countries in the economic and military sectors and the long-term momentum of both infrastructural and educational patterns provide additional reasons for concluding that programs to encourage rural and agricultural development will require a great breadth of vision as well as a willingness to try to transform or redirect a number of patterns and institutions. This must be primarily a task for the peoples and governments involved, but the question remains as to what assistance the industrial countries might provide, were a major effort made in this direction. Before we can fully deal with this question, we must look at the interactions and influence manifested in elite relations between the rich and the poor countries.

[22]See Peter Haggett, *Locational Analysis in Human Geography* (London: Edward Arnold, 1965), and more recently, Brian S. Hoyle, ed., *Spatial Aspects of Development* (New York: Wiley, 1974).

[23]Edward W. Soja, *The Geography of Modernization in Kenya* (Syracuse, N.Y.: Syracuse University Press, 1968).

Elite Relations

In talking about current elite relations between the rich and the poor countries, we are really talking about the interactions between what might be called "the World War II generations" on the side of the rich countries and a mixture of generations on the side of the poor. Perhaps because of the relative stability of regimes in the industrial world, as well as the long apprenticeship period in gaining high position, an extremely high percentage of top elites in the industrial countries went through the social–historical crucible of World War II. Given the lack of any generational studies (in the historical–cultural sense of Marias), one can only suggest that the international scene has been profoundly influenced by what would appear to be the hierarchy of global priorities and sensitivities contained in the world view of this generation: (1) the world is seen primarily in terms of global military-political balances and struggles (often with a frosting of moral righteousness); (2) matters of international economics are of next concern, particularly trade, access to markets, and the various international measures to facilitate these; (3) last (in terms of priorities) comes a variety of social and environmental matters, which it is assumed must often be sacrificed for larger economic, political, and military goals, much as was the case during World War II.

The elites in the poor countries are much more varied, although it is clear that there is a consistent urban bias among them.[24] A great deal depends upon whether their country gained independence in the 19th century (as did many Latin American countries), the early 20th century (Turkey), or after World War II. Among those gaining independence after World War II, much depends upon the manner in which independence was achieved. If it was the result of a bloody struggle (Algeria) rather than a simple transfer of power (much of Africa), then the respective roles of the military and the civilian elites are quite different. Also, the different colonial patterns and practices of the French, British, Dutch, and Americans set different contexts within which the current elites operate.

Military Interactions. Given the fact that there has been a tendency

[24]For a full elaboration of this point and its harmful impact on agriculture and rural development, see Michael Lipton, *Why Poor People Stay Poor: Urban Bias in World Development* (Cambridge, Mass.: Harvard University Press, 1977).

for politicians and historians in the West to ignore the large role of the military in shaping our modern industrial society, it is not surprising that with some exceptions (like Latin America) there has been a similar neglect in the study of the role of the military in the development process of the poor countries.[25] In addition to the way in which independence was achieved, there are a number of factors that influence the contemporary role of the military in the poor countries. Cultural homogeneity or lack thereof can be a major influence in how politicized the military becomes. The kinds of neighbors one has is obviously important. Particularly important are the kinds and degree of contact between Western and non-Western military elites. Historically, these contacts were directly related to colonialism, and indigenous forces were trained primarily to maintain domestic order. They were used only occasionally for border wars or to assist the colonial power in its own wars. Variations in organization and training were found between the French, British, and Dutch, particularly in the respective roles assigned to the military and to the local police or *gendarmerie.* Most senior officers received their training in the military academies of the mother country.[26] Another historical influence was the degree to which the colonial powers were or were not able to build upon indigenous military traditions.[27]

The influence of Western elites upon non-Western military elites today depends not only upon the traditional role and legitimacy (or lack thereof) of the military; it depends greatly upon the interaction and congruence of views regarding regional threats and the international strategic situation. Until recently, sub-Sahara Africa was seen to have relatively little international strategic value. As long as that was the case, it was primarily the former colonial powers that sought to

[25]"There is an extraordinary reluctance . . . to acknowledge how deeply war and preparations for war have molded social organization in this century, and determined technical and industrial progress. However, . . . many of our civilian institutions, ways of thinking, techniques of organization and control, were evolved first in armies or during wars. The Schlieffen Plan preceded the Marshall Plan. Staff colleges preceded business schools. The first schools of engineering and technology were military." Correlli Barnett, "The Education of Military Elites," in *Governing Elites,* edited by Rupert Wilkinson (New York: Oxford University Press, 1969), p. 193.

[26]For a detailed description of these differences in Africa, see John M. Lee, *African Armies and Civil Order* (New York: Praeger, 1969).

[27]See Stephen P. Cohen, *The Indian Army* (Berkeley: University of California Press, 1971), for a discussion of British policies for drawing upon Indian military traditions.

exercise military influence, and this through their ties with the army. There were dramatic exceptions—like the Congo and the Nigerian Civil War—but the few interventions of the superpowers stood out all the more because of the dominant pattern. Compare this pattern with that of South Asia or the Middle East, where the military influence of the former colonial powers has been severely limited, not so much by any reduction in the amounts of their "assistance" as by the sponsoring of huge new military establishments—especially air forces—by the superpowers. The creation of these air forces has not only upset the traditional dominant role of the army (as well as creating serious interservice rivalries), but has changed the dominant external military influence from the former colonial power to one of the superpowers.

The approach of the United States since World War II has been one of providing both training and weapons to those countries seen to have some strategic value. Weapons are provided under a variety of conditions: there are direct grants of weapons, loans provided for the purchase of weapons, and (increasingly) the direct sale of weapons. These are all coordinated by the Department of Defense, and since 1970 a single officer has been in charge of both the Foreign Military Sales Program and the Military Assistance Program.[28] With continuing increases in the price of imported oil, it can be expected that the United States will continue to expand its direct sales programs to help offset balance-of-payments pressures. Even before the current expansion in sales, the United States was clearly the world's leading arms seller and total exporter, having exported more than $50 billion worth of military material in the period 1945–1970 as compared to $16 billion for the rest of the world in the same period.[29] Even so, the Pentagon has stressed the continued need for training programs:

> While military equipment and hardware may deteriorate in time, an understanding of American technology and management techniques, and most importantly, of American culture, carries over long after the foreign national returns to his own country. Such an understanding and appreciation of the United States may be worth far more in the long run than outright grants or sales of military equipment.[30]

---

[28]James Clotfelter, *The Military in American Politics* (New York: Harper & Row, 1973), p. 207.

[29]Ibid.

[30]Ibid., p. 208. Some 10,000 foreign military personnel are trained annually by the United States (at home and abroad). From 1945 to 1969, some 287,000 foreign soldiers were so trained.

The potential developmental role that the military in the developing countries could play has not been clearly explored either in United States practice or in the literature on economic development. One of the underlying reasons has been the lack of clear conception of the role of the military in societal development in general and the lack of appreciation of the particular problems that the military face in the newly (post-World War II) independent countries. While the actual legacy of colonial military practices can be determined only by a country-by-country or region-by-region analysis, one of the key dimensions is the relationship between the political and the military. In many cases, the analysis must include not only relationships at the national level but also the degree to which colonial military law and security arrangements have been revised or not. Without a restructuring of local security arrangements, the fairly common colonial pattern of private armies under the control of district governors tends to persist—with a resultant blurring of distinctions between the political and the military spheres.[31]

The configuration of political-military relations (in their broad sense) will greatly influence the range of possibilities for using the military for development purposes. In India, the role of the military has been carefully circumscribed in a society where, with one recent exception, the political leadership has had high levels of legitimacy. The resulting pattern has been to channel development efforts outside the military sphere. In contrast, in Israel, there has been practically a fusion between the political and the military spheres, with the result that major social and agricultural development projects have been carried out by paramilitary units. The success of these projects has depended upon a high degree of interministerial cooperation (between the ministries of defense, labor, and education) and for this reason has not proved to be a model easily exportable.[32] In much of sub-Sahara Africa, where competition for public employment is perhaps the most critical "policy" question and where successive regimes have low

[31]Lee, p. 182, shows that in sub-Saharan Africa, only Ghana and Tanzania thoroughly revised local security legislation. In India, the several-century battle among British political and military leaders to define their respective roles, made the postindependence decision of Indian leaders to restrict the powers of the army that much easier. See Cohen, pp. 1–31, 170–173.

[32]Lee, p. 135.

levels of legitimacy, the military finds it difficult to remain neutral. Equally, regimes of low legitimacy are less likely to entrust major development projects to the military, as this, in effect, grants patronage privileges to a potentially rival group. Finally, regardless of the particular military-political balances, the generally privileged status (at least economically) of the military is to some degree dependent upon their ties and relationships with military elites in the industrial countries. Any attempt to convert the military to a public works role or into a development agency threatens their external elite linkages much more severely than would be the case with political and economic leaders, should they plunge into genuine development projects.[33]

Political Interactions. The pattern of influence that Western political elites have upon non-Western political elites is similar to that found among military elites. Such background factors as the patterns of socialization and education among non-Western elites, the organizational history and function of particular ministries, the regional position of the country, and the international priorities dominant in that region all shape the nature of the interactions and the relative degrees of influence.

A major difference between the military and the civilian bureaucracies is that the military is normally part of an external reputation system (where the leaders of the military have strong technical, economic, and psychological reasons to be concerned about their international ties and reputation), while the civilian bureaucracies are quite varied as to their relative concern with external versus internal reputations and constituencies. On the one hand, there are ministries that through tradition or function have high levels of external concerns and contacts and are thus concerned with maintaining their external visibility. These normally include the foreign ministry, the planning or development ministry, the ministry for trade or commerce, and often the finance ministry. High officials in these ministries can be expected to travel abroad more frequently, to host visiting officials from other countries and international organizations more often, and to be more conversant with current industrial thinking and technologies than other officials. On the other hand, there are those ministries that are

[33]For a full discussion of possible development roles for the military and the serious practical problems (internal) that are associated with such attempts, see Lee, pp. 132–139.

primarily concerned with internal matters and constituencies; these normally include those handling social and welfare matters, labor problems, education problems (below the university level), and rural and traditional agriculture.

To the degree that a country is dependent upon external markets, financial aid, or technical supplies, it can be expected that the various ministries will use their external contacts in an attempt to improve their domestic bargaining position.[34] Other ways in which the basic bureaucratic bargaining context and power balances can be changed include budget "feasts or famines,"[35] shifts in budget management systems (such as to zero-base budgeting), changes in ministerial decision-making procedures,[36] and the more traditional means of a new government, or even in some cases, what has been called a "ruralizing election."[37] As pointed out earlier, the relative influence of the rich countries in any given case will depend upon the historical development of a particular country's institutions, economy, infrastructure, etc. However, it is clear that any attempt to increase the relative importance of rural and other programs that would be genuinely adapted to the needs of the traditional agricultural sector will have to overcome the momentum of many current international priorities and aid programs that have tended primarily to strengthen the industrial and urban sectors and their associated ministries and bureaucracies. These trends have been additionally strengthened by the activities of multinational corporations, to which we now turn.

Multinational Corporations. Multinational corporations have been a popular topic for study among social scientists in recent years; however, following their general urban bias, there has been much less

[34]That this can also be the case where these contacts are "negative" can be seen in the oft-noted ways in which the Soviet and United States military cooperate in providing mutual threats that strengthen their respective bargaining positions at home.

[35]Graham T. Allison, *Essence of Decision* (Boston: Little, Brown, 1971).

[36]Examples here would include changes like the Freedom of Information Act and, more especially, the "environmental impact statement" procedures under the National Environmental Policy Act.

[37]See Leslie L. Roos, Jr., and Noralon P. Roos, *Managers of Modernization* (Cambridge, Mass.: Harvard University Press, 1971), pp. 202–204. They define such an election as one where "the political power and representation of the rural masses increases; the power of an urban-based, bureaucratic elite declines." Their prime example is Turkey, although they indicate this has happened as well in Sri Lanka, Burma, Jamaica, and Lesotho.

study of the agricultural multinationals. Historically, the pattern is much the same. While there are innumerable studies of the economic influence of foreign corporations in any period one chooses from 1900 on, very few discuss the larger political dimensions, much less anything to do with agriculture. The only exceptions are found in some specialized studies by agricultural economists and in some studies in the Marxist tradition that have sought to look at the imperialistic dimensions of corporate power. The major concern of Marxist scholars in regard to agriculture has tended to be with the large plantation companies, particularly those in Latin America.

In terms of the developmental time-frame, it would seem clear that the potential of the agricultural multinationals for profoundly influencing the nature and structure of Third World societies (for better or for worse) is much greater than that of the industrial multinationals, both because these countries are commonly two-thirds agricultural and because the daily livelihood of the peasant masses and the ultimate viability of the regime rest upon the health and productivity of agriculture. Given such a tremendous potential for influencing this basic sector of society, it is curious that there has been so little serious interest in this area on the part of most social scientists.

From the available data, it would appear that the agribusiness multinationals, as Western institutions, are tending to follow the same pattern of behavior that the colonial countries followed in the 18th and 19th centuries: they are exporting their own agricultural ideas, technologies, and management practices; trying to transpose them directly; and adapting them only when it becomes clear that a particular practice or technology is extremely inappropriate (see Chapter 2). One noticeable difference among agribusiness multinationals relates to their historical background; general corporate situations and strategies vary significantly between those companies with a long history of foreign involvement in agriculture—normally those working one or more plantation crops—and those companies that have tended to enter the market since World War II (primarily companies involved in selling agricultural equipment and supplies as well as those processing food). The plantation-based companies—like United Fruit and Tate & Lyle—are generally vertically integrated, have a fairly wide geographic compass, and have a cushion of political influence built up in the days of the "banana republics" and "gunboat diplomacy." They are in a position to play different countries off against each other, particularly since they tend to operate in small countries that are largely dependent upon a single plantation crop.[38] These conditions tend to buffer them somewhat against the winds of economic nationalism and the major risk that they run, expropriation.

The new agribusiness multinationals that have entered the foreign field since World War II have placed a great emphasis upon agricultural "development" (i.e., the introduction of industrial agriculture). They are generally concerned with medium-sized and larger countries, as they see the greatest market potential there; often these are also the countries upon which international agencies and national governments focus their development aid. As a result, while the old plantation firms tend to have their primary contacts and influence with long-standing ministries and elites (both at home and abroad), the new equipment, supply, and processing firms have more contacts and influence with the newer national planning and development ministries and with the various national and international aid agencies.

At the national level, contacts, projects, and alliances with multinational corporations tend to become a part of the competition between various members of the national elite and between their various bureaucracies and ministries. This competition concerns not only the vested interests of the groups involved but also basic policy questions about the direction of the economy, the balance between private and public sectors, and the short-term benefits of "turnkey" projects where production on needed items can begin quickly, versus the longer-term benefits of developing an indigenous engineering and technological capacity.[39] Also, to the degree that the agribusiness

[38]For a full discussion of the social, political, and economic dimensions of plantation economies, see George L. Beckford, *Persistent Poverty: Underdevelopment in Plantation Economies of the Third World* (New York: Oxford University Press, 1972).

[39]A case study describing the interaction of these different types of competition is found in "The Dilemma of Technological Choice: The Case of the Small Tractor," by G. S. Aurora and Ward Morehouse, *Minerva* 10(October 1974):433–458. The study describes how one of India's national laboratories, the Central Mechanical Engineering Research Institute, sought to develop an indigenously designed and produced small tractor (20 HP), which they named the *Swaraj* ("self-rule" or "independence" tractor). At the same time, the Minister of Industries was promoting local production of a Czech tractor through a collaborative project whereby the Czechs would help build the factory and would supply many of the parts. The ensuing bureaucratic competition involved politicking between the national ministries, the cabinet, and the planning commission;

multinationals become involved, questions emerge regarding the relative priority of agriculture versus industry as well as the type of agricultural development to be fostered. Unfortunately, the latter question has rarely been addressed directly in most developing countries. There has been a gradual shift in national priorities toward a greater sectoral emphasis on agriculture, but the prevailing conception has been that this should be along "modern" industrial lines. Some questions have been raised about the social and political implications of the green revolution, but given the momentum of the Western world views that have been diffused first through colonial educational institutions and then through bilateral aid programs and international development programs, there is little awareness that the long-term energy and environmental costs of industrial agriculture will require even the rich countries to rethink their agricultural policies before long. The multinational corporations, for all their self-proclaimed flexibility and innovativeness, are large, centralized bureaucracies wedded to high-energy technologies and universalized Western concepts of progress. As such, they can be expected to be among the last to adapt on their own part to the requirements of the coming decades. If the activities of multinational corporations, and particularly those in the agribusiness field, are to be of long-term use to a developing country, they need to be carried out in a carefully supervised policy framework. The potential benefits of their activities must be weighed against a whole set of actual and potential dependency relationships that are developing in the agricultural field.

NEW DEPENDENCY RELATIONSHIPS

Several levels of dependence have been suggested in the discussion of the decision-making context of national elites. By examining things in a developmental time-frame, one can see the momentum of infrastructural and institutional patterns established in colonial times.

---

several Punjabi state enterprises and agencies; and the Council of Scientific and Industrial Research. All of the policy questions alluded to above were involved, although in this case, the public-private debate (unlike that involving multinationals) was one of foreign-national public enterprises versus a mixture of local private and local state initiatives.

Attempts to overcome this momentum are complicated by the societal patterns and interests that have grown up around these imports, by generational conflicts, and by the influence of significant external forces. In seeking to overcome some of the dependencies carried over from the past, policy makers in the poor countries ought to be aware of the new kinds of dependence that threaten them and not heedlessly exchange one type of dependence for another.

While I am primarily concerned with outlining the new types of dependency relationships that will predictably result from a widespread dissemination of the green revolution, a brief review of the way in which petroleum dependency (and all of its serious consequences) was allowed to grow up unnoticed may be instructive. *This unnoticed growth in the poor countries reflects perhaps the most serious form of dependence that they face: that of unquestioning acceptance of the basic economic models and concepts of the industrial countries.* Given such acceptance, there was no reason for the elites in the poor countries to think much about their growing petroleum dependency, for the industrial countries themselves were even more dependent and they showed little concern about price rises, embargoes, or shortages. Easy assumptions regarding the long-term availability of cheap energy were shattered in the 1972–1973 period. Massive international dislocations have resulted, and there is considerable turmoil and confusion as countries attempt to reassess the situation.

Another illustrative aspect of petroleum dependency relates to its differential impacts. While many have tried to unfurl a rather large umbrella called *interdependence* to try to show a need for cooperation in facing the harsh storms we find ourselves in, the fact is that the umbrella is tattered and torn—with the result that while some are able to keep themselves tolerably dry, others end up rather damp, and some are drenched. This is clearly the case among the developing countries when it comes to the impact of oil price increases. Four different impacts have been noted[40]:

1. Those countries which will benefit. These are primarily the oil-exporting countries. Some of the larger of these still face serious problems in determining how to distribute the benefits

[40]James P. Grant, "Energy Shock and the Development Prospect," in *The United States and the Developing World,* edited by James W. Howe (New York: Praeger, 1974), pp. 39–41.

among their populations. Examples include Nigeria, Venezuela, Algeria and Indonesia.

2. Those where the impact will be neutral or slightly beneficial. These are countries—like China, Columbia, Mexico, Bolivia, and Tunisia—which are largely self-sufficient in oil and/or have other valuable resources.
3. Those which will suffer disproportionately from any economic slowdown in the industrial countries. Examples here include Mexico, South Korea, Greece, Turkey, Yugoslavia, and Tunisia.
4. Those hardest hit are the forty poorest countries. These are the countries in tropical Africa, south Asia, parts of Central America and the Caribbean, Uruguay, and the Philippines. These countries contain some 900 million people. Their direct increase in foreign exchange costs [for 1974] is estimated to be $3 billion. Indirect costs in terms of loss of agricultural and industrial production, increased costs of imported goods, etc. are probably even greater.

Unequal impacts have also been the rule among the industrial countries and have led to quite divergent approaches regarding the best strategies for the rich consuming countries to follow.

As pointed out in the last chapter, both industrial agriculture and the green revolution "package" are energy intensive and are based on the same faulty assumption that led to current oil difficulties: that cheap energy (and other basic inputs) will continue to be available. There is another more fundamental dimension to the energy problems of industrial agriculture; it depends upon the stock of available energy (those mineral deposits built up over geological time that provide fuel and fertilizers). This is quite in contrast to traditional agriculture, which is based upon the flow of available energy (the daily solar radiation intercepted by the earth)[41]:

> The upshot is that the mechanization of agriculture is a solution which, though inevitable in the present impasse, is anti-economical in the long run. Man's biological existence is made to depend in the future more and more upon the scarcer of the two sources of low entropy.[42]

---

[41]Nicholas Georgescu-Roegen, "Economics and Entropy," *The Ecologist* 2(July 1972):17.
[42]Ibid. Low-entropy energy is that which is easily available and utilizable by man.

Not only are there risks of eventual collapse in continuing to emphasize the scarcer of the two forms of available energy, but (as with other sophisticated technologies) the associated control and distribution mechanisms and bureaucracies tend to be highly centralized and not under the effective control of the ultimate users.

The Role of Agribusiness Multinationals

At a less fundamental level, the activities of the agribusiness multinationals in promoting their farm equipment, supplies, and food-processing techniques threaten the establishment of a new web of dependencies for the poor countries.[43] It must be kept in mind that these companies evolved in the setting of either American or European agriculture. Their underlying assumptions postulate the presence not only of cheap energy but of a sophisticated infrastructure—both physical and institutional—that has evolved to handle temperate zone agriculture. There is, however, some difference in the relative availability of land and labor and in the size of the average farm. In America, equipment and supplies have tended to be larger in size because of the large size of the average farm and because of the scarcity of labor. European equipment—at least up until the "rationalization" policies promoted by the European Communities to reduce the number of small farmers—has tended to be smaller in size and less specialized in function.

From the perspective of the multinational corporations, the introduction of their agricultural wares has been facilitated by several national and international predilections. The key decisions regarding the introduction of foreign farm equipment are often made by people with little direct connection to or interest in agriculture, particularly peasant agriculture. The key decisions often relate to such things as import licenses, foreign exchange rates, permission to buy into domestic firms, etc.[44] Many bilateral aid programs require the purchase of

[43]The activities of the more traditional plantation-associated firms will not be reviewed here. The dependency relationships associated with them are reasonably well known, if not widely disseminated (see footnote 38).

[44]"The government of Pakistan, for example, has encouraged the use of large [foreign] tractors and wheat combine harvesters through its policies of subsidizing interest rates and the price of foreign exchange for imports of agricultural equipment. Even as early as 1964/65, over a third of all loans granted by the Agricultural Development Bank were for purchases of tractors and other mechanical equipment." Keith Griffin, *The Green Revolution: An Economic Analysis* (Geneva: United Nations Research Institute for Social Development, Report No. 72.6, 1972), p. 48.

equipment or goods from the donor. Also aid advisers from abroad are normally most familiar with the practices and equipment prevalent in their own country. The early mystique of the green revolution, the need of policy makers to "do something" in the face of rising populations, the prestige factor associated with having the latest and most sophisticated equipment—these and other ideas regarding "progress" in agriculture made it easier for the agribusiness multinationals to gain access to markets in the developing countries.

At the international level, the agribusiness multinationals have been more successful in their sector than other multinationals in working out comfortable cooperative arrangements with international agencies, especially the Food and Agricultural Organization. This private-sector success in gaining significant access to international development programs is perhaps due to the fact that even in the most free-enterprise-oriented industrial countries, there has been heavy government intervention in agriculture. In the United States, for example, the links between farmers, universities, government, and corporations have been gradually forged over more than a century. If anything, these links and their mutual reinforcement are probably more pervasive in the agricultural sector than those of the military–industrial complex in its sector. Whatever the reasons, the growth of this cooperative arrangement, known as the FAO–Industry Cooperative Program, has been impressive.

Set up in 1965 on the initiative of Herbert Felix, a Swedish food processor, the membership has grown from 18 companies to 84. Although some 19 countries are represented, 23 firms come from the United States and 14 from the United Kingdom.[45] Membership is limited to top-level executives of firms already engaged in agriculture in the Third World, and an annual fee of some $3,000 virtually guarantees that only large firms will be represented. No banks are included. While most companies appear not to have joined with the idea of immediate profits in mind, most "seem to have joined in the long-term hope of gradually convincing the United Nations aid establishment that private investment can play a real part in the development 'mix.' "[46] The Industry Cooperative Program not only has developed a general framework through which industry-generated projects can

[45]Louis Turner, "Multinationals, the United Nations, and Development," *Columbia Journal of World Business* 7 (September–October 1972):13–22.
[46]Ibid., pp. 16–17.

be channeled for FAO and UNDP consideration but has even sponsored (beginning in 1971) some small private "advisory missions" to various governments to give "neutral" advice on agri-industrial planning. Examples of projects in the equipment field include the Massey-Ferguson "School of Agricultural Mechanization" set up in Colombia in cooperation with the FAO.[47]

In addition to establishing closer working relationships with international agencies like the FAO, multinational corporations have sought to develop private multilateral investment banks and development corporations. One such investment bank is ADELA, which focuses its attention on Latin America. It is also a shareholder in the Latin American Agribusiness Development Corporation (LAAD), which was "founded in 1970 to make modest-sized, high-impact investments in the food-supply systems of Latin America."[48] Its founding members include major United States banks as well as many of the same agribusiness multinationals belonging to the Industry Cooperative Program of the FAO. Thus, even before we examine the sectoral impacts of agribusiness multinationals, we can conclude that they have been able to build up an impressive array of information sources in the developing countries, as well as a large number of points of access.

Sectoral Variation

The web of dependencies that the poor countries risk becoming enmeshed in includes the following: dependency upon external sources of energy; external sources of research, technology, and parts; externally controlled distribution systems for vital supplies; external capital and credit; and for their agricultural surpluses (if any), externally controlled shipping, processing, and markets. Even so, any evaluation of the degree of dependency being built up will show

[47]Louis Turner, *Multinational Companies and the Third World* (New York: Hill and Wang, 1973), p. 154.

[48]Ibid., p. 149. "High-impact" should perhaps read "high-profit." On an investment of some $2 million, LAAD earned a net profit of some $500,000 in 1975. This was facilitated by low-interest U.S. AID loans ($17 million by 1976). Most of the projects involved capital and/or energy-intensive specialty crops or products. See Frances Moore Lappé and Joseph Collins, *Food First: Beyond the Myth of Scarcity* (Boston: Houghton Mifflin, 1977), pp. 355–356.

variations between countries, between agricultural sectors, and in the degree of foreign penetration.

In the agricultural equipment field (tractors, combines, plows, irrigation pumps, etc.), the experience with tractors is probably typical. Until recently, the tendency has been simply to export the large four-wheeled models sold in the United States or Europe. Even where there are trained drivers and mechanics, these models meet the needs (and the pocketbook) of only a small percentage of the farmers:

> By not developing models specifically geared to Third World needs, the Western manufacturers left a gap in the market into which the Japanese in particular have moved. In large parts of Southeast Asia, a four-wheeled tractor or large harvester is badly suited to wet and hilly conditions. Firms like Honda have been very successful with two-wheeled, walk-behind machines that in price come much closer to competing with the running costs of the traditional ox and plow.[49]

In spite of these partial adaptations to local needs and conditions, recent increases in petroleum prices have tended to vindicate those agronomists like René Dumont who have stressed the need to devote considerable research and development to improving ox-drawn equipment. And while even conventional agricultural economists have stressed the need for "selective" mechanization, the predictable disruptions of local labor markets engendered by the green revolution have encouraged the large farmer to seek to "protect" his new and expensive inputs from the vagaries of labor availability by mechanizing as much as possible.

In the field of irrigation, the potential dependencies are linked not only to the types of equipment chosen but to the source of water: whether it is from dams, reservoirs, canals, rivers, or groundwater. Critical questions here relate to the nature of the distribution system, who controls it, and its ability to deliver water when and where

[49]Turner, pp. 152–153. The only major United States exception has been Ford, which around 1964 "saw this tractor gap in their product line and approached governments and universities to find out what was needed. Their engineers were then given the task of designing a tractor that would be as simple, cheap, and reliable as possible. Using a seven-horsepower lawn-mower engine, Ford has produced a one-speed, rope-started model that can be easily assembled by local dealers. By 1969 the model was being field-tested in Jamaica, Mexico, and Peru among farmers who used a number of different methods, and in 1970 it went into test marketing in Jamaica. The company intends to sell it only through dealers who are willing to provide fertilizers and agronomic advice as well" (p. 153).

needed.[50] While governments often have a predilection for grandiose projects like the 19th-century canals in India or the 20th-century Aswan Dam, the ultimate usefulness of such projects depends upon a range of decentralized "intelligence" and logistical operations rarely incorporated in either the planning or the running of such large projects. For example, India's "first" green revolution, the development of commercial crops (cotton, indigo, sugar cane), was founded upon the laying out of huge canal systems (based on much smaller European canal systems).[51] Unfortunately, the designs paid insufficient attention to the need for drainage, and often what was in the design was left out during construction or negated by the building of adjacent roads or railroads. The running of the canals was the preserve of the Canal Department, which along with the Revenue and Commerce Departments saw the whole operation purely in terms of the revenue produced from the commercial crops and the canal fees. Writing in 1879, the secretary of the newly formed (and lower-status) Department of Agriculture, Allan O. Hume, argued that this was a "system of spoilation" that risked disaster and that only a series of wide-ranging agricultural reforms (including proper manuring and tillage, small-scale mechanization, credit measures, debt relief, and measures to prevent further salinization) would offer any longer-term remedy.[52] With prescience, he argued further that without such measures,

> come it quickly or come it slowly, the ultimate result here is also certain; and, unless a radical change is effected in existing arrangements, we know, as definitely as we know that the sun will rise tomorrow, that the time must come when some of the richest arable tracts in Northern India will have become howling saline deserts. . . .[53]

---

[50]There would appear to be some interesting analogies between distribution questions in irrigation systems and the recent criticisms of the "trickle-down" theory in economics. If one accepts the idea that economic systems are more analogous to such irrigation distribution systems than to the filtration of water through various natural strata, then it becomes clear that distributive justice in an economic system must be an explicit part of the design of the system and that fairly complicated administrative logistics are involved in ensuring its continual functioning. Otherwise it is highly unlikely that any water will reach the fringe populations.

[51]The following is based on Whitcombe (see footnote 21).

[52]Ibid., pp. 183–185. Proposals for fuel and fodder reserves to provide organic manures were ruled out by the Revenue Department as impractical, while the manufacture of chemical fertilizers from saline–alkali deposits was prevented by the Salt Tax.

[53]Ibid., p. 191. Saline–alkali tracts more than doubled from some .8–1.2 million hectares in 1891 to 2.4–2.8 million hectares in 1971.

India's "second" green revolution "has proceeded on principles not dissimilar to the first: maximization of guaranteed and immediate productivity by means of irrigation-intensive techniques."[54] There appears to be an equal lack of concern with declining soil fertility and with the reclamation of marginal lands. The requirements of the new seeds for water at precise times in their life cycle have put strains upon the century-old roster system for distributing water from the canals. The development of tubewells provides a more assured source of supply, and the Canal Department has promoted the development of state tubewells. Again, however, the major concern has been revenue, and the siting of such state-controlled tubewell stations has been in the immediately productive areas.[55] Private tubewells have also mushroomed, sometimes encouraged by special credit arrangements, but given the lack of adequate groundwater surveys in many areas, there has been an "overcrowding" that has resulted in declining water tables. Consequently, attempts to gain an "assured" supply of water may succeed for only a few years.[56]

It would seem that irrigation-related dependencies are less directly linked to the multinational corporations than is the case in other equipment fields. Rather, the dependencies are tied more to foreign aid and capital (whether bilateral or multinational), since large-scale dams and irrigation projects often require external sources of financing. To the degree that such projects are also tied to export crops, market dependencies increase as well. Each of these measures creates additional external dependencies for the small peasant farmer. While these are often regional or national rather than international, that makes little difference from his perspective. His lack of any influence over regional or national markets, transportation systems, irrigation policies, state or national credit and land policies, and so on means that his meager control over his destiny is reduced even more.

[54]Ibid., p. 192.

[55]Ibid., p. 194.

[56]Two other factors come into play in India. First, crops in India require more water per unit of production because of high levels of percolation. For rice it has been estimated that it takes 3,300 gallons of irrigation water per kilogram in India as compared to 660–880 gallons per kilogram in Japan [Ingrid Palmer, *Science and Agricultural Production* (Geneva: United Nations Research Institute for Social Development, Report No. 72.8, 1972), p. 54]. Second, evaporation losses before the water gets to the fields are extremely high—50–60% from canal to field. The lining of canals and distribution channels with tile or cement reduces these losses but is expensive.

Direct involvement on the part of agribusiness multinationals can be expected in the area of supplying the seeds, and particularly the fertilizers and pesticides, associated with the green revolution. As indicated in Chapter 3, most of the research and plant breeding associated with the green revolution has been carried out by private foundations and various national and international plant research units. Increasingly, it can be expected that the multinationals will become involved in the mass multiplication and distribution of new hybrids.[57] The argument in favor of this runs essentially this way: as the demand for the new varieties increases, only the large multinationals will be able to marshall the necessary capital, extensive facilities, skilled personnel, and quality control. Given the reasonable probability that these multinationals will also seek integrated control over the whole green revolution package (seeds, fertilizers, pesticides, and equipment), there is a real fear on the part of many developing countries that they will be unable to control many of the basic decisions influencing the course of their own agricultural development. Should there be either a vertical integration of seed production back into the research and breeding stage or a horizontal integration over the seed-fertilizer-pesticide package, then

> it may become difficult for developing countries to influence the breeding priorities of foreign private companies whose financial interests will lie in mass production of a few lines rather than to undertake research and development for difficult environments and not so populous countries. While production costs will have been "internationalized" welfare benefits will still be heavily stratified along national and regional lines. There may be increasing difficulty in maintaining plant breeding programmes for difficult environments. It is also by no means certain that the time horizons of private enterprise are adequate to guarantee the preservation of necessary genetic diversity.[58]

In the field of food processing and marketing, there has already been considerable interest and penetration by multinationals. One general approach has been the more traditional one of expanding into the developing countries to obtain various mineral and agricultural

[57]Some are already engaging in their own plant-breeding research as well. CPC International has been developing sorghum varieties in Mexico, and IBEC (International Basic Economy Corporation, a creation of Nelson Rockefeller) has done work on maize varieties in Brazil.

[58]Palmer, *Science and Agricultural Production*, p. 90. The same arguments apply to the very diverse mixtures of fertilizers needed for different regions.

resources that are then processed and sold in the industrial countries. The location of processing is often a bone of contention (e.g., the battles over the location of freeze-dried coffee-processing plants). There does appear to be a tendency—which may increase with higher bulk transportation costs—to do the processing in the developing countries (where cheaper labor is also normally available). In the agricultural field, another major factor is the perishability of the particular crop. With something like tomatoes, it is clear that they must be processed not far from their place of production. H. J. Heinz, for example, has been expanding its research, production techniques, and processing into Australia and Portugal. Such an approach may open up new markets for these countries and may make various research and production techniques available, but it also creates a new set of dependencies based upon far-distant markets and organizations that are very difficult for the poorer (and often even the richer) countries to control.[59]

Another approach has been for the food processors and marketers to seek to expand their sales in the poor countries. The ubiquitous presence of Coca Cola is but one example. At the other end of the nutritional spectrum, a combination of corporate concern and the prospect of contracts from AID-type agencies have led some multinationals to seek ways to improve the protein-deficient diets of the majority of the populations in the poor countries. Three tactics have been employed. First, there have been attempts at the fabrication of new protein-rich products. While Monsanto has had some success with a soybean-based drink called Vitasoy in Southeast Asia (where the taste of soybeans is already well accepted), a number of other such attempts—like the AID-backed project in Tunisia carried out by International Multifoods—have foundered upon people's strong preferences for their traditional diets.[60] A second tactic has been to try to enrich an intermediate product, like flour, which can then be mixed into traditional foods. However, even here, companies have run into taste and texture problems, and some of them have shifted their emphasis from human to animal foods. Third, a few companies have

[59]For a detailed discussion of the operations of H. J. Heinz, see Louis Turner, *Multinational Companies and the Third World,* pp. 155–157. A similar project in Mexico was dropped after several years.

[60]Ibid., p. 158.

tried to enrich the nutritional content of traditional products. Cereal companies in the United States have had long experience in this and, as a result of domestic pressures from consumers and government, have carried out some interesting research. However, the two most extensively tested products—INCAPARINA (developed by Quaker Oats) and "Golden Elbow" macaroni (developed by General Foods)—have been discontinued.[61] While such products would have obvious advantages for those who have the necessary income, they still would fail to reach a large percentage of the poor peasants in the developing countries. As Robert McNamara has pointed out in several of his speeches before the World Bank, we must be particularly concerned about what happens to the poorest 40% of a country's (and the world's) population. This applies to diet and nutrition perhaps even more than to "development."

The nutritional costs to the poor countries of becoming dependent upon foreign markets are perhaps the most insidious. The most obvious facet is the one that Georg Borgstrom has ceaselessly pointed out in his works: that the malnourished countries tend to export badly needed protein to the already-overfed industrial countries.[62] This may take the form of protein-rich anchovies shipped from Peru and Chile to feed cattle and chickens in the United States, or it may take the form of exports of protein-rich groundnuts from Africa in exchange for manufactured goods or high-carbohydrate foods. There are several reasons for this beyond the involvement of agribusiness multinationals. Most important, perhaps, is the urban and economic orientation of most elites in the poor countries. High-protein or other cash-crop exports are often the major source of export earnings that—so the argument goes—must be used to purchase the infrastructure for industrialization. Even if large portions of these earnings were not siphoned off for luxury living among urban elites, it is still not clear that attempts at industrialization should precede attempts at national nutritional self-sufficiency. The latter is not considered a significant goal, not only because of the urban-economic biases of the elites but also because the political pressures upon these elites come from those who are articulate and organized (i.e., other contending urban elites) and not from

[61]Personal communication from Joseph Collins.

[62]See, for example, Georg Borgstrom, *The Hungry Planet* (New York: Collier, 1967), and *Too Many* (New York: Macmillan, 1969).

the poor, the malnourished, and the unorganized. Beyond this, international agencies use economic measures and indicators that rarely include a nutritional factor. For example, according to standard economic indicators, Brazil is booming and Burma and Sri Lanka are stagnating or bankrupt. Nutritionally, Brazil is badly undernourished, while, until recently, Burma and Sri Lanka, with their extensive programs providing free rice rations and subsidized prices for rice, have had generally well-nourished populations.[63] A final complication is that nutritional matters (much like long-term irrigation problems) have a low visibility, which is reduced even more by the nutritional illiteracy of most administrators and economists.[64]

The web of dependencies that can be expected to grow out of a continued expansion of the green revolution in its current form is complex and varies between countries and even within countries. The web is being built, strand by strand, through the creation of a wide variety of extended distribution systems, to and from the farmer. The farmer or peasant is the critical link in the whole system: he is the creative transforming person who changes (with some help from his soil, animals, and plants) all of the various "inputs" into critically needed and life-giving "outputs." Social scientists tend to try to encapsulate such mysteries in what they cheerfully call "black boxes." A more accurate description of the way most planners and researchers perceive the role of the peasant farmer would be that of a milk cow: something that is to be domesticated, kept contented, and fed a diet that it is expected to consume uncomplainingly, all to provide the products that the planners desire. The peasant or farmer has always been dependent in various ways: upon the weather, upon his own and

[63]Ingrid Palmer, *Food and the New Agricultural Technology* (Geneva: United Nations Research Institute for Social Development, Report No. 72.9, 1972), pp. 11, 74–75. The same picture emerges when the new "Physical Quality of Life Indicator" developed by the Overseas Development Council is employed. The PQLI is based on life expectancy, infant mortality, and literacy. Brazil achieved a rating of only 66 (out of a possible 100—the highest achieved by any country), while Sri Lanka achieved an 83. See John W. Sewell, ed., *The United States and World Development: Agenda 1977* (New York: Praeger, 1977), pp. 147–154.

[64]Palmer (ibid., p. 81) has suggested the need for frank discussions between nutritionists and economists on the conflicts between economic growth and a distribution of income conducive to human welfare. Such discussion would clear away much wishful thinking and force "economists to make up their minds on their professional commitment to the human race, a decision they have not as yet been obliged to take."

his family's health, upon the goodwill of the locally powerful, etc. However, it is very doubtful that he is better off trading a slight reduction in these traditional dependencies for a whole new set that is much farther removed from his comprehension and influence and that may risk the loss of his whole way of life rather than his previous periodic losses.[65] It is he—if he is "persuaded" to accept the green revolution package—who must ultimately pay the price of mistakes or miscalculations made by any of the large number of planners, administrators, and researchers who run the various complex systems upon which he becomes dependent. If a planner fails to foresee an oil embargo or a price rise, it is the farmer who pays; if an administrator miscalculates the irrigation rota, it is the farmer who pays; if a researcher does not prescribe the proper insecticide or fertilizer for his local farm ecosystem, it is the farmer who pays. The net result is that the locus of basic farm decisions is moving away from the farmer, and in such a manner as to make it most difficult for him to influence those who increasingly make the decisions. In similar ways, the other regional and national groups, particularly in the poorer countries, also become locked into new dependency relationships.

INTERNATIONAL DEVELOPMENT PROGRAMS

Given the above situation and the professed desire of the international community to assist the poorer countries to "develop," the question remains whether international development programs tend to encourage or to counter these emerging dependencies. Even a cursory review of these programs will show that they do little if anything to counter these trends. Before criticizing the current thrust of some of these programs, it should be pointed out that even the best-designed and best-administered international development programs would face huge problems in trying to counter the momen-

[65]A similar conclusion is reached in the UNRISD report, *The Social and Economic Implications of Large-Scale Introduction of New Varieties of Foodgrain: Summary of Conclusions of a Global Research Project* (Geneva: United Nations Research Institute for Social Development, Report No. 74.1, 1974), pp. 28–29. It is warned that the new technology of the green revolution has a strong tendency "toward a risk fraught reliance on imported inputs, sophisticated research, long-distance transport and world market prices in providing for the daily food supplies of rural and urban populations—in a world not yet free from disorders and catastrophes capable of rupturing the operation of these large-scale systems."

tum generated in other sectors—particularly the trading sector—where the international flows are several orders of magnitude greater. Table 4 illustrates the amount of world trade in primary commodities and manufactured goods and shows the structure of that trade as it is divided up between the rich and poor market-economy countries and the socialist countries.[66]

Table 4. 1974 World Trade in Primary Commodities and Manufactured Goods ($ Billions—FOB)

Exports from: \ Exports to:	Developed market-economy countries	Developing market-economy countries	Centrally planned economy countries	World total
Developed market-economy countries				
Primary commodities	132.3	31.3	8.6	176.0
Manufactured goods	383.8	114.3	28.9	529.0
Developing market-economy countries				
Primary commodities	152.7	39.8	8.5	203.4
Manufactured goods	40.0	14.2	2.1	56.5
Centrally planned economy countries				
Primary commodities	14.8	4.5	10.6	30.1
Manufactured goods	12.9	7.1	31.4	52.1
World total				
Primary commodities	300.4	75.9	27.9	409.5
Manufactured goods	436.7	135.8	62.5	637.6

The figures show that the rich market-economy countries are their own best trading partners in both primary commodities and manufactured goods, while the poor countries have little trade among themselves and are dependent for their export markets upon the rich countries much more than vice versa. This particular pattern has been visible ever since the poor countries became independent, and its momentum is such that the disparities in trade between the rich and the poor countries has been steadily increasing since the 1950s, as can be seen in Table 5.[67]

[66]The data for this table are based on material from the *U.N. Monthly Bulletin of Statistics* 30(August 1976), Special Table C. World totals vary from subtotals because of reexports, differences between FOB and CIF figures, and rounding.

[67]Data from the World Bank Group, *Trends in Developing Countries, 1973,* and presented in Howe, p. 163.

Table 5. Value Indices of Exports, 1950–1970 (1950=100)

Year	From developed to developed	From developed to developing	From developing to developed	From developing to developing
1950	100.0	100.0	100.0	100.0
1955	170.9	147.8	127.6	126.1
1960	243.7	192.9	147.8	132.6
1965	387.4	238.4	195.0	166.5
1970	697.8	370.6	301.0	243.9

In 1974, total world trade was $848.2 billion. This was bolstered by private flows of capital to the developing countries (in the amount of $10.6 billion) for direct investment, reinvestment, and private export credits.[68] These investment figures (and the various vested interests they represent) are barely offset by development aid. Bilateral development aid (concessional) totaled $14.8 billion for 1974, of which OECD countries provided $11.3 billion, OPEC countries $2.2 billion, and the USSR, Eastern Europe, and China $1.3 billion.[69] Multilateral development aid provided by international organizations totaled $6.6 billion.[70] Even this latter figure—representing aid that is presumably not as directly tied to the sort of economic, political, and strategic concerns that so strongly influence the distribution of bilateral aid—must be qualified by the recognition that over 90% of this international aid is channeled through agencies where the rich countries predominate. The World Bank group and the regional development banks all

[68]Roger D. Hansen, ed., *The United States and World Development: Agenda for Action 1976* (New York: Praeger, 1976), p. 192.

[69]Ibid., pp. 203, 211–212.

[70]Ibid., p. 216. The breakdown among the international organizations was as follows (in millions of dollars):

International Bank for Reconstruction and Development	3,168
International Finance Corporation	175
International Development Association	1,138
Inter-American Development Bank	1,064
Asian Development Bank	446
African Development Bank	50
UN Development Program—Special Fund	169
UNICEF, Technical Assistance, Specialized Agencies	138
European Communities	247

have weighted voting systems that reflect the amounts of capital subscribed by the various members, thus giving the rich countries the dominant say in setting priorities. Similarly, the Council of Ministers of the European Communities decides the nature and extent of the aid that it provides to its associate members. While there are few direct political or strategic strings attached to such aid, the conceptions of the rich countries regarding priorities and the proper path to "development" tend to dominate.

Even among the other international agencies (UNDP, UNICEF, the specialized agencies), the position of the major donors guarantees that their concerns are given careful consideration. Additional factors that tend to channel aid from these agencies into conceptual paths similar to those of the World Bank and the bilateral aid programs relate to the background and training of the various staff members who, even if they come from the developing countries, normally have had their training in mainstream industrial economics and planning. Also the various donors have sought to prevent duplication of aid giving. This has resulted in coordinating groups such as the Development Assistance Committee (DAC) of the OECD. Such coordination tends to set the general tone and to spill over into the planning of the UN agencies, the private foundations, and the voluntary relief groups; even so, these latter groups—with their greater flexibility and somewhat different perspectives—have on several significant occasions been able to introduce innovations into official aid programs.[71]

In any case, the resources—whether capital, organizational, or human—available for development aid are quite small when compared to those devoted to trade, capital investment, and military sales and aid.[72] Even within the various development aid categories, the amounts available to international agencies are small, and the amounts outside the various banks (where the rich predominate) even smaller, while the percentages devoted to agricultural aid from whatever

[71]In 1974, private voluntary agencies provided grants totaling over a billion dollars ($1.23 billion), thus becoming a significant element in overseas assistance. Ibid., p. 205.

[72]United States arms sales (roughly 52% of total world arms sales) to developing countries peaked in 1974 at $7,998 million. In addition, $3,075 million in military aid was granted that year. Ibid., pp. 186–187.

source are generally low.[73] The conceptions governing the use of various forms of aid have been largely Western and industrial in orientation. The underlying concepts of "progress" toward industrialization and the "neutrality" of technology (which therefore can be directly "transferred") have been interwoven into almost all aid programs. The increasing emphasis upon agriculture that has been emerging in recent years has raised questions about the high priority previously given to industrialization but has not led to any broad-scale questioning of the assumption of the neutrality and transferability of technology. This has begun to change, with some questioning appearing in studies on ways to generate rural employment, ways to increase the capacity of countries to carry out their own "research and development" (i.e., to be able to adapt research and technology to their own conditions and needs), and in some of the ecological critiques of international development programs.[74]

[73]There has been an increase in recent years, starting about 1966, when there was a shift in aid priorities. U.S. AID assistance for agriculture went from $250 million (or 9% of its budget) in FY 1965 to $504 million (or 20% of its budget) in FY 1967. Much of this and subsequent AID assistance has gone for increased inputs, especially fertilizers. Agricultural loans by the World Bank–IDA have also increased:

Fiscal year	Agricultural loans as a % of total lending	Fiscal year	Rural development loans (agricultural and non-agricultural) as a % of total lending
1948–1960	6		
1961–1965	12		
1966–1970	17	1972	4.9
1971–1972	16	1973	7.5
1973–1974	24	1974	11.0

The main reason for the large difference between "agricultural" and "rural development loans" (which are included in the agricultural totals) is that historically approximately 50% of World Bank–IDA agricultural loans were for large-scale irrigation dams. By FY 1973–1974 this amount had been reduced to 33%, and there was finally some interest on the part of the Bank in seeing that water from such large projects actually reached the small farmer in his field. Sources: OECD, *Aid to Agriculture in Developing Countries* (Paris, 1968), pp. 10, 64; and IBRD, *Rural Development: Sector Policy Paper* Washington, D.C., 1975), pp. 84–88.

The most recent source of funds for agriculture is the International Fund for Agricultural Development, an outgrowth of the World Food Conference. (See footnote 83 for details.)

[74]On rural employment and the need for "intermediate technologies," the ILO report on Kenya is useful. On the need for greater emphasis on local research-and-development capability (as well as the need for developing countries to address the basic problems of the poor), see the *World Plan of Action for the Application of Science and Technology to*

Agricultural Aid

Agricultural aid programs have been influenced not only by the current industrial biases of the wealthy countries but by carry-overs from colonial agricultural policies and research. Agricultural projects during the colonial period tended to be geared toward exports and cash crops. This meant that they were generally established in the best growing areas and that planning and infrastructure were conceived of in terms of large-scale operations. These characteristics and priorities naturally had their influence upon agricultural research. Research tended to be based upon the commercial crops, and those research institutes that were set up tended to be single-crop institutes staffed largely with natural scientists. This pattern has tended to continue as the dominant pattern until very recently, when a few multicrop and regional research institutes have been established. Rarely, however, are any social scientists (other than economists) included in such research institutes. Overall, the emphasis is still on increasing production rather than on rural welfare.

As a result of the colonial experience, administrative structures relating to agriculture have tended to be more bureaucratized and more organized from the top down than was generally the case in the West in the 19th century. In the United States, the pattern was quite the opposite, particularly after the land-grant colleges were established. The network of political, social, research, and educational ties built up between these colleges and the mass of farmers meant that in the United States the organization of agriculture was very much a grass-roots operation until well into the 20th century.[75] Today, agriculture in the West is very much a bureaucratized and centralized operation, so that attempts to "export" current Western experience simply reinforce the old colonial pattern and its neglect of the masses of small peasant farmers. Even without such reinforcement, the previously mentioned

---

Development (New York: United Nations, 1971). Two ecological critiques of current development approaches are Raymond F. Dasmann, John P. Milton, and Peter H. Freeman, *Ecological Principles for Economic Development* (New York: Wiley-Interscience, 1973), and Anthony Vann and Paul Rogers, eds., *Human Ecology and World Development* (New York and London: Plenum Press, 1974). On the ethical dimensions of technology transfer, see Denis Goulet, *The Uncertain Promise: Value Conflicts in Technology Transfer* (New York: IDOC–North America, 1977).

[75]The situation today is quite different, with the network largely dominated by the interests and conceptions of agribusiness.

momentum of colonial education systems still strongly orients local bureaucrats and administrators toward urban values, while the centralized nature of most of their bureaucracies means that advancement and rewards tend to come by pleasing those in regional or state capitals. Table 6 gives one example of the great differences in perceptions found between local administrations (district governors) and local villagers and their headmen.[76]

Table 6. Elite and Mass Perceptions of Villagers' Problems (Turkey, 1962)

Most important problems	District governors (%)	Village headmen (%)	Male villagers (%)
Need for education	29	7	5
Poverty	31	8	10
Need for roads	3	22	20
Need for water	1	27	31
Need for land	8	13	15
Need for occupational equipment	8	4	1
Need for health care	0	0	0
Villagers' own characteristics	6	0	0
No problems	0	2	7
Other	15	5	6
No answer, don't know	0	2	5
Number of respondents	80	424	3,022

The bureaucratization and professionalization of agricultural research and extension work has a number of practical implications. At a personal level, researchers often become more interested in writing professional papers and developing "model" projects than in seeking ways to improve the lot of the many small farmers, who, while they may be grateful, are rarely in a position to provide either sanctions or rewards for what the researcher may or may not do. Traditional status distinctions discriminating against manual labor or modern tastes for comfortable work in a lab may also be sufficient to discourage research oriented toward the small farm. Other factors militating against such an orientation relate more to organizational and political matters. As stressed earlier, agricultural organizations are usually near the bottom

[76]Source: Roos and Roos, *Managers of Modernization*, p. 184.

of the governmental totem pole in most developing countries. This means that neither the financial nor the organizational resources are there to embark upon innovative programs—particularly when attempts to meet the needs of the masses of small peasants would require huge programs. Any such program shift would encounter great political resistance in many countries since existing programs tend to benefit the larger farmers and absentee landlords. Normally, local political power structures are dominated by these same individuals, and any shift to programs that would genuinely benefit the small peasant farmer—whether in the form of guaranteed credit schemes, rationalization of land tenure laws, or marketing cooperatives—would also tend to undermine the network of social, economic, and political sanctions that give the large farmers and creditors their dominant political position. There are no easy answers here, but one point that should be clear is that those development aid programs that seek to improve the condition of the small peasant farmer must necessarily address the question of how existing administrative and political power patterns need to be changed if such programs are to have any chance of success. While aid-providing agencies have their own intellectual and political reasons for avoiding this basic question, failure to address it will likely lead to worse results in the long run than if there were no aid dispensed at all.[77]

The World Food Conference

In spite of the inertia of many current programs and bureaucracies, there may be new pressures building to raise and confront some of the basic international and national political and economic questions relating to development and aid programs. The recent series of ad hoc international conferences—on the environment (Stockholm), on the law of the seas, on population (Bucharest), on food (Rome), and so on—have all tended to raise a host of new issues and problems (as well as to give formal recognition to others long acknowledged). These ad hoc conferences, as well as the recent special sessions of the General

[77]For a disillusioned view of the way in which current United States foreign assistance programs (public and private) are run, see William Paddock and Elizabeth Paddock, *We Don't Know How: An Independent Audit of What They Call Success in Foreign Assistance* (Ames: Iowa State University Press, 1973).

Assembly, have also raised the fundamental question of the distribution of the world's resources and of the relative sharing of the costs and benefits of their development and use.

The World Food Conference in Rome offered a good example of this process.[78] As the global food situation clearly deteriorated in 1973 and as a number of implicit trade-offs between oil-producing and food-producing countries were sensed, there were proposals for a major initiative on food. Although the FAO had held World Food Congresses in 1963 and 1970, there were several reasons that neither the oil producers nor the food producers wanted to have a conference under FAO sponsorship. Generally, the oil countries felt that the FAO was too dominated by the rich industrial countries, particularly in regard to those trade and aid aspects under the control of the Committee on Commodity Problems. The food-producing countries, while preferring to keep agricultural trade negotiations within the confines of the FAO and the GATT, were dissatisfied with FAO bureaucratic rigidity in other areas and, perhaps more importantly, wanted to find a means for including the Soviet Union (not a member of FAO) in the negotiations.[79] The precedent of an ad hoc United Nations conference was most appealing and was eventually adopted.

The World Food Conference was the most quickly prepared UN conference to date, with members having only a year's time to prepare, rather than the usual two or three. In the various preparatory sessions and bargaining, there were basic differences regarding the degree to which agricultural trade should be brought into the purview of the conference, differences regarding whether food problems should be dealt with separately or as part of an overall development scheme, differences over the question of whether the conference should deal primarily with longer-term problems or should also attempt to deal with the possible risk of a major famine in 1975, and differences over the relative responsibility of the rich countries to provide aid and assistance versus the responsibility of the poor to make genuine attempts (including social and economic reforms) to increase their own food production. Also, while there was fairly general consensus re-

[78]See Robert S. Jordan and Thomas G. Weiss, *International Administration and Global Problems: An Analysis of the World Food Conference* (New York: Praeger, 1976), for a full discussion of ad hoc conferences, especially the World Food Conference.

[79]China had just resumed its membership in FAO in 1973, after being absent for 22 years.

garding the need for some sort of international food reserve system, for various means to stabilize international commodity fluctuations, and for some sort of systematic followup machinery to the conference itself, there were wide variations in the proposals submitted to try to achieve these goals.[80]

The conference itself presented a kaleidoscope of contrasts. There were the well-fed delegates from 133 states, delegations from 6 liberation movements, representatives from various UN bodies and specialized agencies, delegates from 28 intergovernmental organizations, and people from international and national nongovernmental organizations. As at the previous ad hoc conferences, there was an iconoclastic daily newspaper (this time called *Pan*) put out by the Stockholm Eco group. They also tried to encourage delegates to pay a "fat tax" for their individual overweights. The hungry were not there, and the representatives of one of the hungriest countries—Bangladesh—eventually walked out in disgust over the failure of the conference to deal with the short-term need in 1975 for food grain aid of some 10 million tons.

The measures that were agreed upon all tended to be oriented toward future rather than immediate action. This is not surprising given the slowness with which international and national leaders have tended to respond to environmental problems of any kind. In addition, food and hunger problems have tended to be compartmentalized, and there has been little work of any kind exploring their multifold interactions with other developmental parameters. Agreement was reached on some general principles and general levels of need—if not all of the recalcitrant details—on the following items.[81] A World Food Council of 36 states was suggested and later approved by the General Assembly.[82] An International Fund for Agricultural Development based on

[80]See particularly the documents considered at the Third Session of the Preparatory Conference, UN Economic and Social Council, E/CONF.65/3, E/CONF.65/4, and the report of that session, E/CONF.65/6.

[81]UN Center for Economic and Social Information, "Outline of Solution to World Food Problems Agreed on by World Food Conference," OPI/CESI NOTE FOOD/13, 19 November 1974.

[82]The World Food Council is located in Rome and has the services and expertise of the FAO available to it; however, it has sought to be its own master in drawing together the various facets of food and hunger policy from the many agencies and departments of the UN and national governments. The Executive Director is John Hannah, former U.S. AID director.

voluntary contributions from the old and new rich was agreed upon, although no specific targets were set beyond one of $5 billion by 1980 for agricultural assistance (from all sources).[83] In regard to food aid, it was agreed that 10 million tons of cereals plus other food commodities should be provided with firm commitments made three years in advance.[84] There was also an "international undertaking on world food security," which sought establishment of a world grain reserve that would serve as a buffer stock and provide a system of world food security. Neither the size, the nature of control, nor the mechanisms for allocating and distributing these food reserves was settled.[85] There was also agreement on the need to improve the data, reporting, and forecasting of food reserves and weather and on the need to encourage an "early warning system" for famine; however, there have been no indications that China and the USSR will cooperate in providing crop estimates. Finally, the conference passed a draft Universal Declaration on the Eradication of Hunger and Malnutrition and a number of resolutions covering diverse areas. The two areas where no agreement was reached were the already-noted failure to arrange for 10 million tons of cereal food aid for the first half of 1975 and the persistently difficult area of agricultural trade. All that could be agreed upon there was to continue to pursue negotiations with GATT and UNCTAD.

Did the World Food Conference represent a new beginning or the last hurrah of a system about to collapse? Looking back, we can see that

[83]After more than two years of hard work and bargaining, the International Fund for Agricultural Development is now in operation. It is interesting in that its $1 billion in funding comes primarily from the OECD and OPEC countries. It has a tripartite structure, including the OECD and OPEC countries (the donors) and the developing countries (the recipients).

[84]This goal has yet to be reached. Food aid shipments in 1975–1976 totaled 6.9 million tons, those for 1976–1977 totaled 8.3 million tons, and those for 1977–1978 are expected to total 9.6 million tons. Although consensus was reached on a 500,000-ton International Emergency Reserve, allocations to it for 1976 were only 92,500 tons. By February 1978 this reserve increased to 436,500 tons. Three-year advance commitments have yet to be achieved—in part, because of reluctance on the part of the U.S. Congress.

[85]At the May 1977 preparatory meeting for the third session of the World Food Council, the secretariat urged that the favorable crop conditions of 1976–1977 be used to establish at least a 20-million-ton reserve of wheat and rice. However, this will be difficult to achieve, given important differences between the United States and the European Communities over the size, acquisition, cost, and release of such reserves.

it was probably a bit of both, although mostly it demonstrated the momentum of the past. It opened several new avenues by its attempt—for the first time at the international level—to trace the many interrelationships among farming, ecology, nutrition, trade, the role of women, economic power over land, the disproportionate costs of weapons systems, and so on. However, the painfully slow progress since the conference in trying to develop any interrelated policies shows the rigidity of the old and fragmented system.[86] The conference, perhaps because of its ad hoc nature, demonstrated a new sensitivity to problems that had been ignored before and suggested a number of worthy goals. However, the "solutions" proposed to deal with these problems and to reach such goals represent primarily an attempt to apply "more of the same" in terms of providing more short-term Western technical solutions—solutions that are antiecological in the longer term, are based upon the availability of cheap energy, and assume that greater centralization, both nationally and internationally, is the only alternative to chaos and collapse. Grass-roots approaches to rural development, appropriate technologies, and decentralized research and extension services received little discussion. Most importantly, there was no recognition that the rich developed countries will have to rethink their own highly energy-intensive agricultural systems; consequently, the traditional assumption that aid, assistance, and progress are things that flow in one direction from the wealthy industrial countries to the poorer countries still tended to predominate the silent agenda.

[86]The five areas of action identified by the World Food Council at its third session are (1) to increase food production in the developing countries, (2) to improve and ensure world food security, (3) to expand and improve food aid programs, (4) to improve human nutrition, and (5) to liberalize and improve food trade. Only items 2 and 4 represent genuinely new agenda items at the international level. However, item 1, the original goal of the Rockefeller Mexican program, still is given the top priority: "The World Food Council—like the World Food Conference and the General Assembly which created the Council—believes that the key to the solution of the world's food problems is to increase food production in the developing countries, particularly in those that are food deficit countries. Enough food to feed all hungry or malnourished people is the first requirement. Oratorical speeches or well-written essays will not feed hungry people." Statement by Dr. John A. Hannah, Executive Director of the World Food Council at the 34th Session of the Economic and Social Commission for Asia and the Pacific (ESCAP), 8 March 1978, p. 3.

SUMMARY

In trying to understand the momentum of the past as it bears upon national decision-makers, as well as trying to understand the increasing difficulties we face because of environmental pressures—such as increasing populations, changing climate patterns, soil erosion, and social disintegration—we can make a simple analogy. The leaders of the world (political and intellectual) are very much in the position of someone on top of a huge snowball rolling downhill. There are two basic problems such a person faces. First, since the snowball picks up more snow as it rolls, he had to run faster just to stay on top of the snowball. Second, he is facing away from the direction he is traveling. Most political and intellectual leaders are well aware that they are running faster and faster just to stay on top of things. Very few are aware that they are facing backward and that to try to give the snowball some direction they have to do the even more difficult thing of turning their heads (and thinking) around to face the real future ahead.

5

New Approaches to the Future

Population must increase rapidly, more rapidly than in former times—are ere long the most valuable of all arts will be the art of deriving a comfortable subsistence from the smallest area of soil. No community whose every member possesses this art, can ever be the victim of oppression in any of its forms. Such community will be alike independent of crowned kings, money kings, and land kings.

Abraham Lincoln, *1859*

In restating the Jeffersonian ideal, Lincoln cast his eye and his logic toward the future and its pressures. His suggestions incorporated a concern both for longer-term practicality and for social justice. Indeed, he argued that the one could not be achieved without the other.[1] The criticisms of the green revolution and the analysis of the momentum of current policies have suggested the need for an analogous but new look to the future that will consider the need for social justice as well as longer-term practicalities. This task is unfortunately much more difficult than it was in Lincoln's day. To try to understand fully how the present has evolved out of the past, a new approach—contextual analysis—has been developed and applied in outline form. To try to understand the various future limitations, options, and alternatives in

[1]See the full text from which the above is taken: "Annual Address before the Wisconsin State Agricultural Society, at Milwaukee, Wisconsin, September 30, 1859" contained in Roy P. Basler, ed., *Abraham Lincoln: His Speeches and Writings* (New York: World Publishing, 1946), pp. 493–504.

agriculture, this same new approach will be applied, again in outline form.

It will be recalled that the argument used in applying contextual analysis to the past (especially the device of the three time-frames) was that only by understanding the full context within which policy is made—that is, its real setting in space and time—could one make a full evaluation of it. Several analogies were used to clarify both the need and the utility of applying the three time-frames simultaneously. One analogy was that of different-scale maps. It was suggested that to understand the spatial dimensions of a city, a city map was not sufficient; one also needed to understand its setting on state and world maps. Only then would the details on the city map take on full meaning. No one of the maps is sufficient by itself to give a full picture because of the inherent trade-off between detail and comprehensiveness at the different scales. This analogy also suggested the vital importance and necessity of employing quite different units of analysis when one examines long-term, medium-term, or short-term historical trends and events.

A second analogy was employed to get at a more fundamental problem, that of change over time. Here it was suggested that not only do the developmental "plays" differ as one follows another, but that the evolutionary "theater" also changes over time. The role of individual "actors" is thus different between plays and may even be different between acts of the same play (the generational phenomenon). In trying to trace both the appropriate units of analysis and the types of major change in each of the three time-frames, a number of points that have generally been left out of the calculations and conceptions of most social scientists became clear. By focusing primarily upon the policy time-frame, most social scientists have tended to regard as constants processes that over real time may be changing significantly. The population explosion—the only one of these longer-term processes that has been noted by any significant number of social scientists—has generated as much intellectual confusion as light. Other significant longer-term changes were noted—such as those relating to climate, deforestation, soil erosion, and genetic simplification—but as yet these are engaging the interest of only minute numbers of social scientists.

In addition to the tendency to project current conditions and trends into both the past and future, the tendency to cast current

Western technologies and conceptions upon the rest of the world was noted. It was suggested that this was and is inappropriate since Western science and technology are an integral part of modern industrial society—which is urban-oriented, highly specialized, highly centralized, and capital- and energy-intensive—things that hardly fit the needs of poor, decentralized, agricultural countries. The dilemma involved in trying to suggest new approaches to the future revolves around the question of how to simultaneously include a genuine concern for longer-term processes and costs while overcoming our cultural ethnocentrism regarding the superiority of industrial society.

What is suggested here is that we need to engage in an intellectual process similar to the attempt made above to understand the full historical context of policy by examining the past in terms of evolutionary and developmental time-frames. The argument in brief is: to make intelligent policy for the future, we must first ascertain those broad evolutionary limitations, parameters and goals to which we must be sensitive if we are to survive as a species; next, we must examine the limitations and parameters facing us in the next century and we must establish our goals for this long period, *keeping in mind that these goals must be consistent with evolutionary limitations and goals.* Only then can we turn to specific policy recommendations for the relative short term of the coming decade. Rather than trying to deal with those problems that may appear important to us based on current experience and practice, framing our decisions on trade-offs between obvious short-term costs and benefits (anything longer than a decade being very long-term from the perspective of most decision makers), I will try to spell out more systematically the evolutionary and developmental limitations and goals that must be consciously included in decision making if it is to promote either survival or social justice. Here again, it is hoped that the employment of the three time-frames as heuristic devices will facilitate the discussion and will show that alternative agricultural strategies for the future can best be evaluated only by including these longer-term considerations.

EVOLUTIONARY PARAMETERS, LIMITATIONS, AND GOALS

The fact that we need to talk about the evolutionary capacities of the global ecosphere as they relate to the survival prospects of the human species may well be the most damning criticism of modern industrial man that can be made. Not only does it suggest the need to try to analyze the causes for modern man's failure to live within the ecological constraints of his environment; it also suggests that we must try to avoid any rose-colored tints being added to our picture of the future as a result of using the lenses that got us where we are. It is here that the previous analysis of the various biases of modern science and thought—particularly as they are linked to specialization—is useful. It has been scientific, technological, and institutional specialization that have facilitated the (uneven) spread of industrial society, while making most difficult any overall or holistic assessment of their impact and likely future costs.

Two great human evolutionary trends—the "explosion" of industrial society and the associated explosion of population—have had increasingly serious impacts on the natural environment. Attempts at assessing these impacts have been made difficult by the momentum of past scientific thought and the division and mapping out of the scientific world in terms of social, physical, and life sciences. Thus, many interdisciplinary studies involve the useful but limited (from an evolutionary perspective) task of showing how one specialty has neglected important phenomena that can be provided by bringing in materials from another specialty and incorporating or synthesizing them. Even those works aimed at a broader overview have been hampered by the lack of holistic models, and especially by the fact that materials must be drawn largely from specialized studies. These specialized studies have all been developed through a process of winnowing out those materials not historically deemed useful in trying to obtain precise knowledge of that limited area. For example, it is mainly since World War II that there have been attempts to examine how man's activities might influence the climate and how his burgeoning numbers relate to global water supplies, land availability, etc.

While the need for better (i.e., more holistic) science is critical, it may be that from an evolutionary perspective the most that it can offer is a bridge from industrial society to some sort of postindustrial society that, like many traditional societies, is able to live in harmony with the environment. To the degree that science is inextricably linked to modern industrial society, one would expect basic changes in science as industrial society is either transcended or collapses. As with the larger transformations that will be required of society if collapse is to be avoided (see below), science will need to become less specialized and less centralized. Before examining these points, the risks of collapse need to be assessed; such an assessment involves trying to judge whether current evolutionary trends are changing the parameters of the global ecosystem and what this means in terms of mankind's general evolutionary limitations.

In trying to assess the evolutionary limitations of the global ecosystem, much less whether its parameters are changing, one is struck full force with the limitations of available data. Available surveys all suffer from the effects of specialization, the traditional gap between the natural and the social sciences, the lack of historical data with a global baseline, and the lack of evolutionary models linking human activity and physical changes in the environment.[2] This is not to criticize them, for they have tried to bring together those materials that are available; however, it is important to be aware of the gaps and weaknesses because these are the things most in need of future elaboration or modification.

The Hydrologic Cycle

When one looks at the global availability of air, water, and soil (past, present, and future), all of the above difficulties become strikingly apparent. Since in an evolutionary time-frame all of the above interact and may be modified by cumulative human actions, new theories as well as new data are needed. In terms of natural phenomena, the hydrologic cycle perhaps best illustrates the interactions between air, water, soil, and climate (see Figure 4).[3] Both natural

[2]A representative sample of these surveys would include: Robert R. Doane, *World Balance Sheet* (New York: Harper, 1957); William L. Thomas, Jr., ed., *Man's Role in Changing the Face of the Earth* (Chicago: University of Chicago Press, 1956), 2 vols.; Report of the Study of Critical Environmental Problems (SCEP), *Man's Impact on the Global Environment* (Cambridge, Mass.: MIT Press, 1970); Paul R. Ehrlich, Anne H. Ehrlich, and John P. Holdren, *Ecoscience: Population Resources Environment* (San Francisco: Freeman, 1977); Report of the Study of Man's Impact on Climate (SMIC), *Inadvertent Climate Modification* (Cambridge, Mass.: MIT Press, 1971); National Academy of Sciences Committee on Geological Sciences, *The Earth and Human Affairs* (San Francisco: Canfield Press, 1972); UNESCO, *Status and Trends of Research in Hydrology, 1965–74* (Paris: UNESCO, 1972).

[3]From Georg Borgstrom, *Too Many: A Study of the Earth's Biological Limitations* (New York: Macmillan, 1969), p. 133.

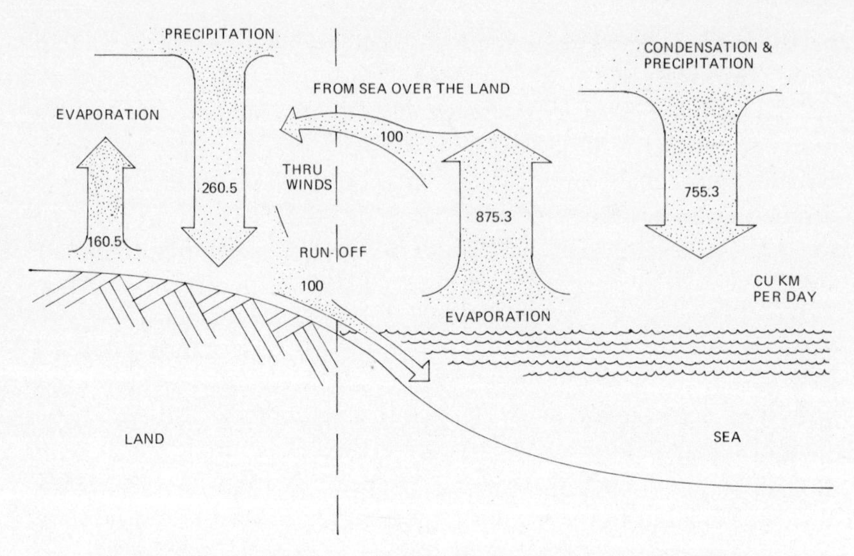

Figure 4. The hydrologic cycle of the Earth (in cubic kilometers/day).

and human processes can modify the relative rates and distribution patterns of the hydrologic cycle. Natural changes may result from changes in climate, which in turn may be due to geological, biological, or solar changes. Human-induced changes may be traced to traditional activities such as deforestation or to more modern ones such as the generation of high-altitude particulate matter.[4] What is lacking is not only an evolutionary history of the hydrologic cycle and man's impact upon it but any data that would give rates of global soil erosion, rates of global deforestation, rates of groundwater depletion, etc. Research carried out during the International Hydrologic Decade, plus the development of satellites and other monitoring equipment, has given us fairly good baseline data on current rates of flux between the four basic water "reservoirs" (see Figure 5).[5] The steps needed next are the development of baseline data for earlier periods and the incorporation into present physical models of the relative impact of various kinds of human activity (see below). Such attempts will suffer from the traditional division between natural and social history, from the Eurocentric nature of much of the historical data, and from the lack of interest

[4]See the SCEP and SMIC reports (footnote 2).
[5]*Inadvertent Climate Modification* (SMIC), pp. 96–99.

on the part of the social scientists (particularly economists) in such "free" resources as air and water.

The potential use of the hydrologic cycle (redefined to include human activities) as an evolutionary indicator linking soil, air, water, climate, and human activities of global scope is so great that it deserves detailed examination here, and hopefully elsewhere. What then are some of the evolutionary problems, parameters, and limitations that are suggested by such an indicator? First, it should be noted that water—in very much the same manner as energy—is critically essential to all aspects of life, human and nonhuman. It appears that we will soon be hit with a water crisis very much like the energy crisis. Even more than with energy, which was never considered a "free" resource, but rather a fairly cheap and available one, any water crisis will be most difficult to deal with conceptually, as illustrated below. For human purposes, the availability of fresh water, particularly for agriculture, is crucial. However, the global availability of fresh water is limited and is subject to losses either through salinization or through a cooling trend in climate that would lock up additional fresh water in polar ice or

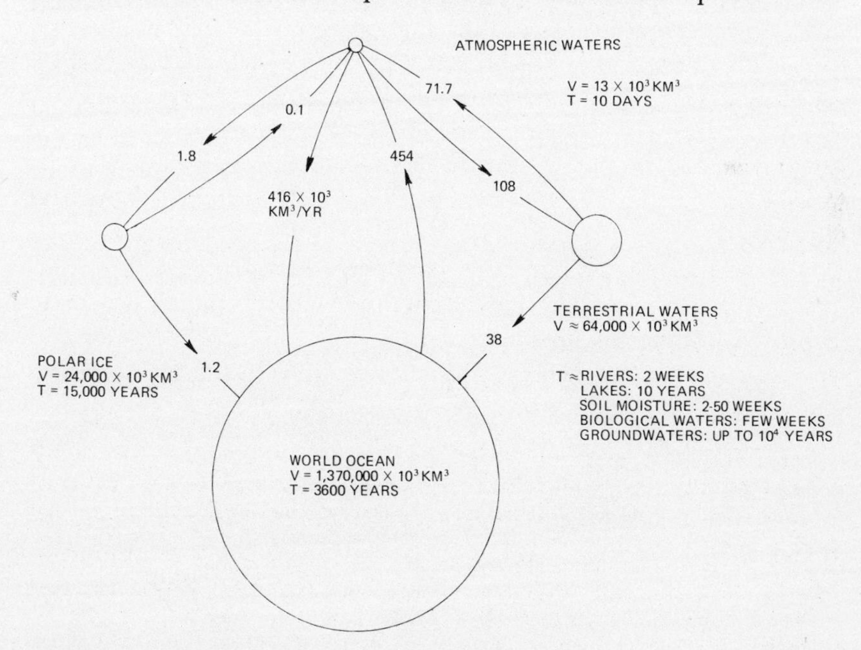

Figure 5. Principal statistical and dynamical characteristics of the hydrosphere.

glaciers. It should be noted that 97% of the earth's water is seawater and that of the remaining fresh water (3%), two-thirds of it is in polar ice and glaciers.[6] Thus, only 1% of the earth's total water is fresh and is available to man.[7]

The global distribution of fresh water through rainfall is mediated by topography, plant cover, climatic patterns, etc. On a global basis, there appear to be a number of feedback mechanisms between various parts of the cycle. Deforestation—which can result from forest fires or tree diseases, but primarily from man's activities—can trigger soil erosion as well as local climate changes. Little is known about regional or global deforestation—either in terms of rates or impact.[8] Typically, Europe has been better studied, and we know that forests expanded somewhat after the collapse of the Roman Empire to where at least 80% of central Europe north of the Alps was forested in 900 A.D. By 1900, this area had only 24% forest cover.[9] The causes of this deforestation were complex and included the assault of the early church on the forests, both to stamp out the paganism associated with them and to develop agriculture. Later, changes in agricultural technology (which encouraged larger fields) and the increasing rate of industrialization put even greater pressure on forests. The impact of industrialization—with its demands for both fuel and construction materials (especially for the sailing ships needed for both trade and warfare)— was great. Even so, the overall rate of deforestation in Europe appears to have been fairly gradual and to have occurred without serious soil erosion.

Deforestation rates on other continents have varied; when combined with less favorable climatic conditions, deforestation has led to serious soil erosion. China has had a long history of deforestation and soil erosion, which reached serious proportions in the 20th century.

[6]National Academy of Sciences, *Productive Agriculture and a Quality Environment,* Washington, D.C., 1974, p. 138.

[7]As in the energy field, many "simple" technological fixes have been proposed—such as cloud seeding, nuclear desalinization plants, or the melting of polar ice. None of these survive any close examination of ecosystemic ramifications or cost. See Chapters 5–9 in Borgstrom for detailed criticisms.

[8]The most recent attempt at an overall assessment is that of Erik P. Eckholm, *Losing Ground: Environmental Stress and World Food Prospects* (New York: Norton, 1976).

[9]National Academy of Sciences, *The Earth and Human Affairs* (San Francisco: Canfield Press, 1972), pp. 52–54.

The new regime—ironically one pledged to materialism rather than the verbal respect for nature of earlier ones—has made major efforts at reforestation. The impact of different regimes has also been visible in Japan. In the early Meiji years, forests were destroyed through taxation, with disastrous flooding the result. Thereafter, model forest protection laws were instituted and maintained until the United States' occupation in 1945. Then again, briefly, occupation authorities sought to expand agricultural land at the expense of forests until the resulting floods led to a reinstitution of the forest protection laws.[10]

Large parts of Africa and Southwest Asia have been reduced to semideserts by the grazing of goats and livestock.[11] Overgrazing was the major factor in the ruin of much of Spain's agricultural land in the 15th and 16th centuries.[12] The exportation of these and other deforestation practices through the Spanish Conquest led to the loss of huge acreages in Latin America, especially Mexico. In North America, pressures to expand agricultural and grazing land, compounded later by the timber industry, led to a much more rapid rate of deforestation than in Europe. Few of the impenetrable forests that Tocqueville marvelled at in Michigan remain, although evidently there is some natural reforestation occurring throughout the eastern United States.[13] Current pressures on forests in Africa, Asia, and Latin America result from population pressures, the search for more agricultural and grazing land, and the continued use of wood for fuel, plus the pressures of industrialization.[14]

The critical relationship between the type of ground cover, water run-off rates, and soil erosion can be seen in Table 7.[15] It should be pointed out that the building up of the topsoil necessary for agriculture (seven to eight inches) requires between 2,000 and 7,000 years.[16] While

[10]Borgstrom, pp. 1–2.

[11]*Inadvertent Climate Modification* (SMIC), p. 63.

[12]*The Earth and Human Affairs,* p. 55.

[13]On Tocqueville's journeys, see Alexis de Toqueville, *Journey to America* (New Haven, Conn.: Yale University Press, 1959). In the 19th century, as much as 80% of New England's forest cover was removed. Today up to 80% has returned. *The Earth and Human Affairs,* p. 54.

[14]For a detailed description, see Eckholm.

[15]Data drawn from Jean Dorst, *Before Nature Dies* (Baltimore: Penguin Books, 1971), p. 134.

[16]Ibid.

Table 7. Rates of Topsoil Removal (in Years)

Ohio:	Time to remove 7–8″ of topsoil	
	Forested areas	174,000
	Meadows	29,000
	Rotated cropland	100
	Corn monoculture	15
Congo:	Time to remove 6″ topsoil	
	Forested areas	40,000
	Grassland	10,000
	Cotton fields	20

there was a period of great national and international concern with soil erosion in the 1930s—intimately linked to the Dust Bowl—the establishment and domestic successes of the Soil Conservation Service, plus our general faith in our ability to manage and control the environment, have lessened our concern with global rates.[17] Even so, it is estimated that some 3–4 billion tons of soil are lost in the continental United States each year.[18] In addition to losses from soil erosion, there are those due to the conversion of agricultural land to other uses: roads, subdivisions, airports, and other types of industrial development. For example, in the state of California over 3% of the total agricultural land (some 3.5 million acres) has been converted to other uses.[19] As one of our most valuable resources—one that can be maintained through proper care but that when lost can be replaced only in thousands of years—it is clear that there is need both for national and global inventories and for a whole new range of social, legal, and economic measures to ensure the preservation of soils.[20]

Beyond the pressures of agriculture and grazing upon forests, and

[17]The classic studies were those of Hugh H. Bennett, *Soil Conservation* (New York: McGraw-Hill, 1939), and Graham V. Jacks and Robert O. Whyte, *The Rape of the Earth: A World Survey of Soil Erosion* (London: Faber and Faber, 1939).

[18]*Productive Agriculture and a Quality Environment,* p. 146.

[19]*The Earth and Human Affairs,* p. 100. For the United States as a whole, "during the past 20 years some 11 million hectares (an area larger than Ohio) have been converted into urban areas and highways." David Pimentel, William Dritschilo, John Krummel, and John Kutzman, "Energy and Land Constraints in Food Protein Production," *Science* 190(21 November 1975):760.

[20]For example, the normally conservative Michigan Department of Agriculture, concerned about loss of good agricultural land through conversion, has proposed a whole new set of property rights and relationships for farmland—essentially a system of joint ownership by the state and the farmer, with the state ensuring its continued use as agricultural land.

the pressures of population and industrialization upon agricultural land, there are significant pressures upon available supplies of fresh water. To the degree that deforestation continues and is combined with poor agricultural practices, available rainwater becomes runoff rather than being stored in plants, moss, or soils or percolated down to replenish groundwaters. Another kind of pressure results from the fact that water demand for urban areas and industrial processes is much less seasonal than either the agricultural demand or rainfall patterns themselves. The resulting "peak" demand periods are compounded by the geographic distribution of rainfall. For example, in California, while four-fifths of the rain comes in the winter season, peak demand is from May to October; also two-thirds of the rain falls in northern California, while three-quarters of the demand is in southern California.[21]

On a global scale, the mining of groundwater reserves built up over geologic time has reached such proportions that it may be causing a slight increase in the mean ocean level.[22] The loss of such reserves is critical in several respects. In the temperate zone, where evaporation is much less than in the tropics and where the greatest amount of food is produced, the outtake of groundwater is two to three times the replenishment rate.[23] Much of the demand in these areas is for industry, not for agriculture. However, in the arid zones of the tropics, there is pressure to increase the amount of irrigated land in spite of the fact that evaporation rates make groundwater much more costly to use and often unavailable in time of drought. The amount of water required per acre in the Sahara for conventional irrigation is 5–10 times as much as that required in Morocco and 30–50 times as much as that required in southern Italy.[24] Borgstrom has argued that groundwater reserves (rarely adequately surveyed prior to exploitation) should be seen in terms analogous to food reserves—to be used only when there is a shortage: "In harsh hydrological terms this can almost be formulated

[21]Borgstrom, p. 168. Also, as one moves to permanent rather than seasonal irrigation problems of salinization, disease vectors and hardpan buildup are greatly increased. Ibid., p. 191.

[22]*Inadvertent Climate Modification*, p. 183.

[23]Borgstrom, p. 144, estimates the outtake rates as three times the replenishment rate for Europe and two times for North America.

[24]Ibid., p. 193. These differences have spurred interest in various traditional and new ways to conserve water, reduce irrigation, etc. See National Academy of Sciences, *More Water for Arid Lands*, Washington, D.C., 1974.

to mean that groundwater reserves *never* should be used for regular crop production but be held in abeyance for drought relief in critical times."[25] It is rather curious that the World Food Conference has talked about conserving or storing for time of need only the results of agricultural production rather than the water that makes production possible. Perhaps more than with other aspects of agriculture, the global and regional distributional effects and differences in space and time of water use have been ignored.

A final aspect of the hydrologic cycle: the long-term and large-scale activities of man in altering the face of the earth clearly have local and regional effects upon climate and may well have global effects. To the degree that climate is so modified, it may seriously alter the availability of water, change its pattern of distribution, change temperature regimes, etc.—any of which can significantly modify agriculture production. Such man-caused changes may either reinforce or counter naturally occurring variations in climate. While little is known regarding the impact of global changes upon specific regions, it is clear that significant regional heat transfers may result:

> Less easy to appraise are those numerous activities, including such things as irrigation and deforestation, which change the apportionment of heat release from the surface between sensible heat and latent heat of vaporization. . . . Sensible heat is immediately available to drive atmospheric motions. The latent heat becomes available ultimately only when the vapor condenses and precipitates. This usually occurs at an appreciably different altitude and longitude from the place of evaporation, and frequently at a different latitude. Comparatively large amounts of heat are involved in these transfers and their importance should be studied.[26]

In spite of all of the complexities of the hydrologic cycle (particularly when man's influence is added), its careful examination has the virtue of forcing one away from the typical "through-put" approach of economics and agriculture to a recognition that there is a global cycle of an essentially fixed amount of water and that both the quality of that water (particularly fresh water) and its distribution are of critical importance to man's evolutionary future.[27]

[25]Ibid., p. 143.

[26]*Inadvertent Climate Modification*, p. 179.

[27]In some ways the shift in thinking called for here has parallels in the development of the idea of the cycle itself. Yi-Fu Tuan, *The Hydrologic Cycle and the Wisdom of God* (Toronto: University of Toronto Press, 1968), shows that earlier classical versions were challenged in the late 17th and early 18th centuries by a series of natural philosophers of religious bent. In particular, John Ray, in his work *The Wisdom of God*, sought a unifying

Intimately associated with the hydrologic cycle is the cycling of various elements, including six basic ones:

> Of the many elements necessary for life, six rank as the most important: carbon, hydrogen, oxygen, nitrogen, sulfur, and phosphorus. The first three are combined in energy-rich materials such as carbohydrates and oils. These together with nitrogen and sulfur, are essential ingredients in all proteins. The sixth, phosphorus, is needed for the transfer of chemical energy within protoplasm, whether this energy is used for activity (respiration) or growth.[28]

Each of these cycles exhibits wide variation in its distribution and concentration as well as the variable flux rates noted in conjunction with the hydrologic cycle. Some of them have been more intensively studied than others. Less well known are the various interactions between them, much less the impact of man's activities upon them. The complexity of the nonhuman interactions are in themselves staggering:

> Each of these six elements circulates through air, land, sea, and living systems in a vast biogeochemical cycle. The circulation of water (hydrologic cycle), the very slow erosion and uplift of continents (geologic cycle) and the opposing processes of photosynthesis and respiration (ecologic cycle) are all involved.[29]

All of these cycles ultimately intersect in living matter and are subject to the transformations therein as well as to the variegated distribution patterns of different organisms. The impact of modern man on these cycles appears to vary: little impact has been shown upon the oxygen cycle, while there have been significant modifications in the carbon cycle (in the form of CO_2).[30] It has been suggested that all of the major cycles of elements be monitored and that given the essentially nonrenewable nature of phosphorus reserves, an international agency be established "to advise on the prudent production, distribution and use of the phosphate resources of the world."[31]

---

concept of God's bounty as exemplified by the various forms and uses of water. His overarching theological concept eventually led to a number of scientific inquiries and discoveries.

[28]*Man in the Living Environment*, p. 45.

[29]Ibid.

[30]*Man's Impact on the Global Environment* (SCEP Report), pp. 40–99.

[31]*Man in the Living Environment*, p. 59. The only "renewable" sources of phosphates are the guano from fish-eating birds, various plants, and the return to the soil of animal and human manures—something that goes against various cultural, health, and engineering biases.

The same recommendation could, of course, be made for the numerous other mineral resources becoming increasingly scarce. The obvious example of oil suggests the political and economic difficulties involved, particularly at the international level, should there be any serious attempt to deal with such scarcities. The recommendation also suggests two different levels of awareness: economists and politicians are now very aware of at least some of the problems associated with scarcity of those resources that fuel our modern *industrial society;* however, only a relatively small segment of society is aware of the much greater risks associated with the scarcity, squandering, or spoiling of those six elements essential for *life.* It is precisely because of the relatively much greater awareness of the significance of scarcity of industrially useful resources that they will not be discussed in detail here.[32] Another neglected but critical area is the maintenance of the diversity of organisms, plants, animals, and humans necessary for the continued survival and evolution of the global ecosystem.

Maintaining the Diversity of the Global Ecosystem

The risks of genetically simplifying the major crop seeds upon which man is dependent have been discussed (Chapter 3); it is an even more difficult and complex task to assess the evolutionary risks of overall ecosystem simplification. The role of microorganisms in maintaining the global ecosystem is significant, if not fully understood. The oxygen-producing phytoplankton in the oceans have received some study, as have various microorganisms in the soil. Some scientists have expressed the fear that these and other critical microorganisms may be destroyed or suffer mutations because of pesticides or radioactivity. There is increasing evidence that pesticides often do more harm to useful insects than to harmful pests.[33] The loss of a single insect

[32]There are a number of inventories of mineral resources and their expected lifetimes, computed on the basis of different assumptions regarding growth of demand, new discoveries, levels of technology, etc. See Preston E. Cloud, ed., *Resources and Man* (San Francisco: Freeman, 1969) and Hans H. Landsberg, Leonard L. Fischman, and Joseph L. Fisher, *Resources in America's Future* (Baltimore: Johns Hopkins Press, 1963). The persistence of technological optimism in the face of scarcity projections can be seen in the strong reaction, especially by economists, to *The Limits to Growth* (Donella Meadows, Dennis L. Meadows, Jørgen Randers, and William W. Behrens III [New York: Universe Books, 1972]).

[33]See the discussion in Raymond F. Dasmann, John P. Milton, and Peter H. Freeman, *Ecological Principles for Economic Development* (New York: Wiley, 1973), pp. 151–163.

species—the honeybee—would result in incalculable evolutionary and economic losses. The losses of plant diversity caused through deforestation are significant, even if poorly chronicled. The causes of deforestation—agricultural pressures multiplied by population pressures—themselves force simplification of plant systems other than forests. No one knows what the requisite diversity of organisms, insects, plants, and animals is for the global ecosystem to maintain the adaptive capacity necessary for its evolution and survival. However, there are some specific examples of simplified ecocystems that are instructive.

Most notable—particularly given the rates of their decimation—are the tropical rain forests. While remarkably stable, the removal of large acreages results in an irreversible change so that any replacement plant communities are less complex and of quite a different character.[34] The removal of rain forests may change regional rainfall patterns and may result in serious soil erosion—as in Brazil. Other examples of the potential instability of relatively simple ecosystems are to be found in islands. The introduction of foreign animals, plants, insects, and birds has been particularly disruptive ever since the development of modern sailing. The introduction of rats, goats, and rabbits has caused great disruption in islands ranging from Mauritius to Hawaii to New Zealand.[35]

The loss of animal and bird species has been gradually increasing under the pressure of man's direct and indirect attacks upon them. As many have noted, it may be well to consider the loss of these species a sort of distant early warning analogous to the way in which miners traditionally took canaries into the mine with them to warn of the buildup of dangerous gases. While the data are rough and probably somewhat skewed by the historical interests of zoologists, it has been estimated that since the 16th century:

> No less than 120 forms of mammal and about 150 bird forms have vanished. It is estimated that about ten forms (species and subspecies) of birds became extinct before 1700: about 20 in the 18th century: about 20 from 1800 to 1850: about 50 between 1851 and 1900 and another 50 since 1901.[36]

These figures do not include the losses of marine species. While there are some data on the loss of various species of commercial fish (very

[34]See Dorst, pp. 135–145.
[35]Ibid., pp. 218–257. Dorst also discusses imported plant pests, such as the water hyacinth, and imported mollusks.
[36]Ibid., p. 35.

discouraging data), there is little information on the loss of diversity in tidal marsh lands. The latter are especially critical in terms of genetic diversity, not only because they are probably the most productive of all ecosystems but because they are the breeding grounds for such a range of species.[37]

Discussions of whether there is or is not a requisite level of genetic and cultural diversity for continued human adaptation tend to suffer from both theoretical confusion and ideological distortion. It would appear, as suggested briefly in Chapter 4, that some of these confusions could be sorted out by trying to identify the appropriate units of analysis for the different time-frames. As far as human genetic diversity, there appear to be two opposing trends that have both accelerated in the past century. On the one hand, the increased mobility of people in the last century has greatly increased the amount of interbreeding between relatively distinct gene pools. This presumably may lead to greater genetic diversity. On the other hand, the use of modern medicine to save people with various genetic-related health defects has put into the gene pools various mutations that otherwise would have been naturally selected against. While the ethical pros and cons can be debated, it is clear that much like certain domesticated animals and plants, these individuals and their progeny will require continued and significant levels of human intervention and social support to survive.

Questions relating to the value or utility of cultural diversity are equally clouded. First, it is not clear what the appropriate units of analysis are for the evolutionary time-frame. One possible unit is civilizations, even recognizing its value connotations. In his work, Toynbee recognizes some 34 civilizations and divides them into independent, satellite, and abortive civilizations. He is concerned with the actual historical relationships, degrees of affiliation, and ability of

[37]The same applies to the ocean shores, if to a lesser extent. As Thor Heyerdahl has observed ("How Vulnerable Is the Ocean?" in *Who Speaks for Earth?*, edited by Maurice F. Strong [New York: Norton, 1973], pp. 61–62): "Particularly in the eastern and southern Mediterranean and on several of the local islands the yellowish rock is becoming discolored from yellow to gray and even black in a belt six to eight feet wide. . . . The once-yellow rock seems like coke, and in many places sizable clots of oil are hammered into the pores like sheets of tarmac. And this is happening in what is the cradle for the bulk of marine life, since most species have to pass one stage of their life cycle on the rocky shores of islands and continents."

civilizations to adapt or respond to challenges.[38] More theoretically oriented is the work of Sahlins and Service, which develops a theory of cultural evolution.[39] In spite of the sophistication of both these works, there appears to be no consistent use of the key terms and no consideration of whether their content is actually being changed as different time-frames are implicitly applied.

Compounding the difficulties, even if consistent and agreed-upon units could be developed, is the confusion regarding what is meant by *complexity* and *diversity* in the social realm. While we know that in natural ecosystems greater diversity is highly correlated with stability, we do not know if there is a similar relationship in the social realm. Additionally, we tend to confuse diversity and complexity. Diversity is essentially a systemic measure that indicates the number of distinct units (or species) in a system, their relative importance, and their distribution.[40] Complexity is generally a characteristic attributed to a unit within a system. A university may be a complex institution, but that tells us little about the diversity of the society of which it is a part. Equally, while each tribe in Africa in the 17th century may have appeared relatively simple in social structure, there was tremendous diversity among the tribes on the continent. Today, the various African states may each appear more complex, but there has been a loss of diversity. The same point could be made for other continents, and the question is whether the spread of a dominant industrial-technological pattern around the globe—with a resultant loss in cultural diversity—threatens social stability in the same way that it threatens the global ecosystem through simplification.

The policy implications that result from the above discussion of evolutionary parameters and limitations are necessarily general in nature. They would appear to revolve around a simple maxim: *Let us keep our evolutionary options open.* To close out any such options at a time of increasing environmental and social stress would appear to be most foolish. While there is great need to pursue additional scientific knowl-

[38]Arnold Toynbee (in collaboration with Jane Caplan), *A Study of History* (London: Oxford University Press in association with Thames and Hudson, 1972). See p. 72 for his chart on the various civilizations and their relationships.

[39]Marshall D. Sahlins and Elman R. Service, eds., *Evolution and Culture* (Ann Arbor: University of Michigan Press, 1960).

[40]Eugene P. Odum, *Fundamentals of Ecology* (Philadelphia: Saunders, 1971), pp. 143–154.

edge regarding the various evolutionary limits and the critical aspects of such things as the hydrologic cycle, global deforestation and soil erosion rates, and genetic variability, the even greater need is for restraint or regulation of the dominant evolutionary trends of population growth and the spread of industrial society. Both of these put extreme pressures upon the global ecosystem and upon social stability; while we are trying to devise strategies to limit their growth, we must constantly seek to preserve what genetic and cultural diversity we have and to encourage an expansion of diversity rather than simply letting the current trend of simplification continue. To use a legal analogy, we must understand evolutionary limitations and parameters in a "constitutional" sense; that is, we must realize that these aspects of the earth and of life are the basic structural elements that constitute the world as we know it. Those changes that are influenced or controlled by man should be introduced or made only after extraordinary consideration, debate, and review. For a change, perhaps the natural scientists should review and consider borrowing from political-constitutional theory a system of thought and practice that sorts out different levels of change as to their relative importance (constitutional, legal, administrative) and provides quite different procedures for dealing with each level.[41] More difficult, however, will be the task of educating and persuading society of the need to exercise extraordinary caution in a number of critical areas.

DEVELOPMENTAL PARAMETERS, LIMITATIONS, AND GOALS

In shifting to the developmental time-frame, one shifts both the span and the focus of analysis. Developmental limitations, parameters, and goals often manifest themselves as regional or national aspects of the various evolutionary dimensions discussed above; also other processes and problems come into focus. It should be recalled that the argument for determining future alternatives using contextual analysis is that policy options must be consistent with developmental

[41]For an elaboration of this argument to include the point that the legal burden of proof should be shifted to those proposing projects that offer a threat to our global environmental "constitution," see Kenneth A. Dahlberg, "A New Approach to Evaluating Longer-Term Energy Risks," *Geothermal Energy* 4(August 1976):39–41.

parameters, which in turn must be consistent with evolutionary limitations. This is rather different than current modes of analysis used in futurology—where one tends to get either extrapolations from current trends or some form of utopianism where alternative societies are more or less simply postulated. Much debate regarding the future and how to understand and deal with it has centered around the computer studies sponsored by the Club of Rome: *The Limits to Growth* and, more recently, *Mankind at the Turning Point.*[42] By using contextual analysis to discuss them, perhaps new light can be shed upon some basic methodological questions.

Both works use computer models with multiple feedback loops. Both also are "real-time" models that use actual historical and resource data. Such models require explicit statements of assumptions and allow projections to be made holding one or more of the variables constant. Both models appear to indicate that "solving" one major "problem," such as population growth, is not sufficient to stem the momentum toward disaster coming from other areas. The models have been very useful in stressing the need for simultaneous strategies to deal with the difficulties found in several sectors. Where the reports differ is on the question of whether one understands the coming century better in terms of a homogeneous global system or as a world still strongly influenced by the momentum of differential rates of growth according to region. Given the discussion in Chapter 4, it should be clear that the latter approach appears much more realistic. However, it is also at this point that a number of the structural weaknesses of the Mesarovic–Pestel model appear, weaknesses that certainly apply to other computer models because they reflect general theoretical weaknesses in the sciences as well as huge gaps in relevant data.

[42]See Meadows, and Mihajlo Mesarovic and Eduard Pestel, *Mankind at the Turning Point* (New York: Dutton, 1974). One of the more detailed—but sometimes misleading—critiques of *The Limits to Growth* was prepared by the Science Policy Research Unit at the University of Sussex. It is found in *Futures* 5(April 1973). The most recent report submitted to the Club of Rome is not a computer study and consists of sectoral policy recommendations made by 21 international experts. See Jan Tinbergen (coordinator), *RIO: Reshaping the International Order* (New York: Dutton, 1976). For a useful review of other types of environmental modeling, see The Holcomb Research Institute, *Environmental Modeling and Decision Making: The United States Experience* (New York: Praeger, 1976).

The importance of the structure of a computer model is recognized in both works. *The Limits to Growth* stresses that

> the basis of the method is the recognition that the *structure* of any system—the many circular, interlocking, sometimes time-delayed relationships among its components—is often just as important in determining its behavior as the individual components themselves.[43]

Mesarovic and Pestel point out that

> the conclusions drawn from the analysis of the world future development depend on the view of the world as embodied in the structure of the computer model.[44]

Several levels of caution are needed. First, verbal descriptions of the structure of the model and its world view may or may not correspond to its logical and mathematical structure. Second, while both works talk about environmental and ecological variables, it would appear that virtually none of the variables discussed above in regard to the evolutionary time-frame have been incorporated. How, for example, would climate change, global soil-erosion rates, or depletion of fresh water be dealt with? In part, this lack may result from the fact that both teams are more strongly oriented toward the social than the natural sciences. It also results from the general lack of awareness and data on such matters. A third weakness relates to questions of differential boundaries and qualitative change.

Differential boundaries are incorporated formally in the "valves"—that is, the equations that give either a negative or a positive multiplier (and its rate) to a particular feedback loop. There are serious data questions regarding the upper and lower limits within which such multipliers operate in the real world. Variable boundaries, however, also are linked to the "units" chosen, and if the units themselves are transformed in the operation of the system, then how are the separate equations to be maintained? For example, all of these questions would be involved in trying to construct an accurate computer model of the making of bread. During the process, your initial "units" of flour, sugar, yeast, etc., become transformed into dough. The rising of the dough operates within fairly narrow temperature limits that we know empirically—but that would be very difficult to estimate purely on a biological or a chemical basis. During the process of baking, there is

[43]Meadows, p. 31.
[44]Mesarovic and Pestel, p. 54.

another transformation, and again it occurs within fairly narrow temperature and time limits. The basic point here is not to suggest that an accurate computer model could not be made of the process—this could be done since most of the basic descriptive data are available, particularly those for the transformations. The point is, if you know nothing about the making of bread dough, and your model essentially describes various mixtures of flour, sugar, yeast, etc., the model is not capable of forecasting or suggesting a basic transformation of the system and its structure.[45]

To apply this specifically to the Mesarovic–Pestel model: while the model does include the structural possibility of major shifts between and within the 10 regions around which it is structured, it does not include the possibility of transformation to a system of either 5 or 15 regions (this could be done by structuring new models but cannot result from the current one). Equally important, the 10 regions chosen are essentially political-economic regions, with perhaps some cultural shadings. One has only to list other possible regions to suggest the variety of ways in which regions can be conceptualized and the multiple ways in which they differentially interact upon one another: cultural regions, climatic regions, regions as defined by major river basins, geographic regions, regions according to vegetation, soil regions, etc. The regions chosen, while probably adequate for many of the purposes of the Mesarovic–Pestel report, still reflect a social science orientation. While difficult to deal with and to try to synthesize, greater discussion of the interactions of natural and social systems is clearly warranted.

The above criticisms suggest that while the two models may be useful in demonstrating the complexities of current structures and trends, the models themselves are incapable of dealing with the question of how the structure of the system may be modified (either intentionally or unintentionally). The authors of both studies, in effect, make their own judgments and recommendations as to what sort of changes will lead to a stable output for their model. *The Limits to Growth* recommends a policy of dynamic equilibrium where population and capital are constant in size and other inputs and outputs are

[45]Another example is smog. Prior to the analysis and description of the synergistic photochemical reactions involved in its creation, it would have been impossible to develop a computer model that would have projected its creation out of CO_2, NO_x, etc.

minimized. The *structural* changes necessary to achieve this are not specified; rather, there are discussions about the need for individuals to change their values. *Mankind at the Turning Point* talks about the need for organic growth—that is, balanced and differentiated growth—in order to get away from the disasters implied in current exponential growth curves. While they do talk about the need to redistribute wealth among the 10 regions of their world model, little is said about how to do this other than persuading leaders that any other approach is more costly.

Structural Changes Needed in Industrial Countries

What appears to be needed—particularly in the developmental time-frame—is an analysis of the basic *structural* shifts needed in society to keep our evolutionary options open. While such shifts logically should be discussed for each region since their evolutionary and developmental histories vary, as do their future problems, the analysis here will be of Western society because it offers the greatest global threat to keeping evolutionary options open. In the chapter on agricultural alternatives there will be a parallel but more detailed discussion of the structural needs of the poor regions. The basic question to be pursued here is: What are the basic structural changes needed to make Western society more adaptive and less of a threat to global diversity? It will be suggested that we need to move away from the current overemphasis of four basic characteristics of Western society and to turn toward a new synthesis. These four hyperdeveloped characteristics are (1) dependence upon energy-intensive technologies; (2) overindustrialization; (3) overcentralization; and (4) overspecialization.

Western society is currently involved in the painful process of trying to adjust to an energy shock. While the Arab oil embargo simply highlighted the increasing energy scarcity that various experts had previously cataloged, and while it came early enough to give us a wake-up jab rather than a knockout punch, the shock has been great. A good part of this shock relates more to social-psychological phenomena than to the minimal readjustments that have been made in energy-consumption patterns so far. Any time there is a significant gap between a prevailing belief (a myth) and reality, there are two ways to try to bring the two into closer correspondence. One is the very painful psychological process of reevaluating your belief—which becomes all the more difficult if it is a central belief, because other associated parts of your world view will also have to be reevaluated. The other approach is to try to restructure reality so that it fits more closely with your original belief. The former process is all the more complicated because normally a number of vested interests grow up around the original belief, adding economic costs to the social and psychological difficulties of any reassessment. Examples of the two approaches can be found in several areas. On racial matters, there is on one hand a serious attempt to reevaluate the whole range of black history and on the other the attempt of many Northerners to "re-create" the imagined security of their essentially white childhood by fleeing to newly created suburbs. In foreign policy matters, there is the contrast between those reassessing the dogmas of anti-Communism and their corresponding military postures and those—generally with strong economic links to the military–industrial complex—who proclaim that the only way toward "détente" is through military strength.[46] In the energy field, the contrast is clearest in the debate over nuclear power and particularly over the development of breeder reactors. While a number of independent scientists have argued for a moratorium while a full-scale reassessment is made of energy needs and alternatives, the nuclear industry continues to argue that we can and should restructure reality to fit their belief in the easy availability of cheap energy. It is not only the impressive array of vested interests—linked into a fairly tight complex—that impells them in that direction but also the great psychological difficulty of seriously questioning their whole complex of values, particularly those of technological optimism and the equation of progress with industrial growth.[47]

While we are clearly entering a period of systemic readjustment in the energy field (as well as an intellectual reassessment of priorities and values), it is not clear whether the shorter-term adjustments will be more in the direction of further energy intensification and centralization or whether there will be significant shifts toward more diverse and less energy-intensive systems. Although the momentum is clearly in favor of the former, there are a number of developments that

[46]For a full discussion of this approach and how it applies to the "Communist myth," see John H. Kautsky, *Communism and the Politics of Development: Persistent Myths and Changing Behavior* (New York: Wiley, 1968).

[47]Similar difficulties probably help explain the virulent attacks by conventional economists upon works such as *The Limits of Growth.*

suggest an increasing reassessment. At the international level, the increase in oil prices has led to a serious reconsideration of the role of multinational corporations, to the raising of serious questions regarding the green revolution, to questions of energy conservation as regional policy, etc. At the national level (in the United States), there is not only debate between Congress and the president as to what is the best energy policy but the quiet building up of a literature—both theoretical and empirical—that seeks to measure energy efficiencies and to reassess the performance of various sectors of the economy and various modes of transportation in these terms rather than in dollar terms.[48] While widespread use of energy measures would represent a great advance over dollar measures, ideally one should have a series of indicators to cover other types of physical and environmental costs (relating to water, air, and soil) as well as social costs.

In the longer term, regardless of the particular indicators used, the basic need is to restructure our energy systems and technology along "softer" paths so that they are less energy intensive.[49] One example of a policy that would encourage this approach and would also encourage decentralization (see below) would be to discourage current profligacy by setting up a national system of graduated energy rates.[50] This would encourage some efficiency in consumer use of energy and even more in production. Such *framework legislation* aims at an overall net result, while still allowing flexibility within the outer constraints. Such a system would encourage the building of smaller, gas-economizing autos while at the same time making bus and train travel relatively more attractive. Even so, if someone wanted to pay the price, they could still choose to have a large, gas-guzzling auto—although they would probably have to give up other high-energy options since the

[48]The literature here is constantly expanding. For a sample of the various approaches, see Howard T. Odum, *Environment, Power and Society* (New York: Wiley, 1971); Carol Steinhart and John Steinhart, *Energy: Sources, Use and Role in Human Affairs* (North Scituate, Mass.: Duxbury Press, 1974); Robert A. Herendeen, *An Energy Input–Output Matrix for the United States, 1963: User's Guide* (Urbana: University of Illinois, Center for Advanced Computation, Document No. 69, 4 March 1973); and Wilson Clark, *Energy for Survival* (Garden City, N.Y.: Anchor Books, 1975).

[49]For an excellent overview of why this is necessary and why we must rapidly move in this direction, see Amory Lovins, "Energy Strategy: The Road Not Taken?," *Foreign Affairs* 55(October 1976):66–96.

[50]For details, see Kenneth A. Dahlberg, "Towards a Policy of Zero Energy Growth," *The Ecologist* 3(September 1973):338–341.

price differential between small and large cars would be so much greater. Various phases could be built into such legislation to cushion some of the transition difficulties, and there could even be regional differentials to account for resource availability or climatic variations. There are many other possibilities, but the basic problem lies in trying to educate and persuade people to shift from ad hoc short-term remedial measures to framework approaches that will gradually generate momentum in the needed directions.

The shift away from overdependence upon energy-intensive or "hard" technologies is directly linked to the need to decentralize and restructure industrial-agricultural balances. At a theoretical level, this need results from the basic ecological relationship between the energy efficiency of ecosystems and their simplicity or diversity. The more diverse an ecosystem, the greater its energy efficiency.[51] In order for modern industrial society—which is complex but not diverse—to become more energy efficient, it will have to become more diverse. To explain how this may be encouraged, it may be helpful to disaggregate some of the ideas, interests, and ideologies that are associated with industrial centralization.

The basic operating principles of industry since the 19th century have been the use of functional specialization and the division of labor in order to create production of scale ("mass production"). As Tocqueville noted when the industrial system was beginning to emerge, there are two great dangers in such a system. First, the working man is weakened, alienated, and dehumanized, for "in proportion as the principle of the division of labor is more extensively applied, the workman becomes more weak, more narrow-minded, and more dependent. The art advances, the artisan recedes."[52] Second, he feared that extensive industrial development would lead to the creation of a "manufacturing aristocracy," which in many ways would be harsher than any previous aristocracy.[53] While modern governments have grown and expanded their areas of jurisdiction in order to try to deal with some of the abuses flowing from this particular mode of production, they have done so at a cost: they too have had to become more

[51]Ramón Margalef, *Perspectives in Ecological Theory* (Chicago: University of Chicago Press, 1968), pp. 20–23.

[52]Alexis de Tocqueville, *Democracy in America* (New York: Vintage Books, 1954), Vol. 2, p. 169.

[53]Ibid., pp. 168–171.

specialized, more bureaucratized, and more standardized in order to try to regulate industry.[54]

It is important to point out the political consequences of large-scale industrialization because it would appear that many so-called economies of scale in reality result from the ability of the large corporation, through its political and economic leverage, to "externalize" a number of social and environmental costs. For example, the steel, auto, and paper industries have historically externalized great environmental costs by polluting the surrounding air and water. Also, the social costs of importing huge work forces to urban manufacturing centers have been borne primarily by the cities and the states—especially in times when large numbers of workers are laid off. However, since economists have traditionally seen air and water as "free" resources and have had little in the way of interest or methodologies to deal with the larger social and political costs of various industrial enterprises, such practices have not been effectively noted or challenged. Also, once a large industry is established, it is that much more difficult to regulate—again because of its political and economic leverage. The spread of industrial abuses to the national and international level make regulation even more difficult. In a very real sense, the Japanese have mortgaged the health of present and future generations in order to maintain a competitive edge over other manufacturers; equally, large multinational corporations have shifted production to areas where social and environmental legislation and regulation are minimal.

As indicated above, large-scale industrial centralization is built upon functional specialization. Each unit in modern industrial society is complex, but the system is a simple system because all the units are essentially similar. Such simple but specialized systems are not only less energy efficient than more diverse systems but also less stable (i.e., less adaptable). For example, if a serious oil shortage had hit the United States in the 1920s, there would have been a much greater chance for a flexible, adaptive response because of the diversity both within the auto industry (large and small firms; gas, steam, and electric cars) and

[54]There are, of course, other reasons for the growth of "big government," ranging from the impact of World Wars I and II upon government to natural "Parkinsonian" tendencies. For a detailed analysis of the tension between political democracy and industrial aristocracy, see Michael D. Reagan, *The Managed Economy* (New York: Oxford University Press, 1963).

between the different sectors of the transportation industry. Railroads, trolleys, buses, interurban lines—all could have been expanded with relative ease while whatever necessary shifts took place within the auto industry itself. Today, however, not only do we have three and one-half large firms of the same mold, but we have structured our whole life-style as well as the urban-suburban landscape around the automobile.[55]

The gradual establishment of nationwide, centralized, but simple infrastructural systems has been fairly common in the United States. Not only is this pattern found in the energy sector, in transportation, and in industry, but it also appears to be increasingly the case in communications and in agriculture. The process by which this has occurred in agriculture is particularly interesting both because of the many basic incommensurabilities between agriculture and industry and because of the 19th-century and early-20th-century attempts to establish, through the land-grant colleges and the extension service, a series of locally adapted research-service networks. The expansion of industrial agriculture in the past 25 years has been impressive. As in the automobile industry earlier, large numbers of farm equipment firms have been gradually absorbed into a smaller number of large firms. Much the same has happened with fertilizer, pesticide, and irrigation suppliers. Over this period, there has also been a shift in farm policy—first unconscious, then deliberate—from policies favoring or protecting the small farmer to policies favoring the big industrial farmer. A combination of the migration of small farmers to urban areas, the expansion of the suburbs into prime farmland, the decline of small towns, the gradual loss of political power by farmers—particularly at the state level, as reapportionment decisions removed the earlier overrepresentation of rural interests, etc.—all have reinforced the trends toward largeness in both farms and agribusiness. During this same time, the focus of the land-grant colleges also gradually shifted from primary concern with the problems of the small farmer to an interest in all of the scientific, chemical, and mechanical aspects of industrial agriculture. Strong and mutually reinforcing links

[55]For a prescient series of predictions (made in 1958) on the eventual impact of the interstate highway program, see Lewis Mumford, "The Highway and the City," in *Environment and Society*, edited by Robert T. Roelofs, Joseph N. Crowley, and Donald L. Hardesty (Englewood Cliffs, N.J.: Prentice-Hall, 1974), pp. 208–219.

have been forged between the colleges, the large agribusiness firms, the Department of Agriculture, and Congress.[56]

Simultaneously, policy regarding agriculture has come to be dominated more and more by the thinking of economists, whether in government, industry, or university departments. Given the many incommensurabilities between industry and agriculture, it is not surprising that economists have tended to take their existing conceptual framework—which is closely linked to the growth of industry—and apply it in a procrustean manner to agriculture, ignoring difficulties or trying to do away with them by restructuring farm landscapes and populations through powerful technologies and policy measures.[57] The resulting view of agriculture is both partial and distorted—something that, combined with the instability of a simple centralized system, makes policy formulation for and within agriculture most difficult. For example, the unexamined assumption that cheap energy would be available indefinitely has already led to severe national and international dislocations. In another area, the centralization of hybrid crop research and production has generated a number of risks. Major crop losses to wheat rust and maize blight resulting from genetic simplification and monoculture have already been discussed (Chapter 3). In addition, given the four- to five-year lead time required to get new varieties into full crop production, assumptions regarding the constancy of climate become critical, particularly as they relate to the length of the growing season. If there continues to be a shortening of

[56]For a detailed description of these links, see Jim Hightower and Susan DeMarco, *Hard Tomatoes, Hard Times: The Failure of the Land Grant College Complex* (Cambridge; Mass.: Schenkman, 1972). For a good history of the development of the land-grant colleges, see Earle D. Ross, *Democracy's College* (New York: Arno Press and The New York Times, 1969) (reprint of the 1942 Iowa State University edition).

[57]As pointed out by Ernst F. Schumacher, *Small Is Beautiful* (New York: Harper Torchbooks, 1973), p. 47, economists lump together in one category four fundamentally different categories of "goods."

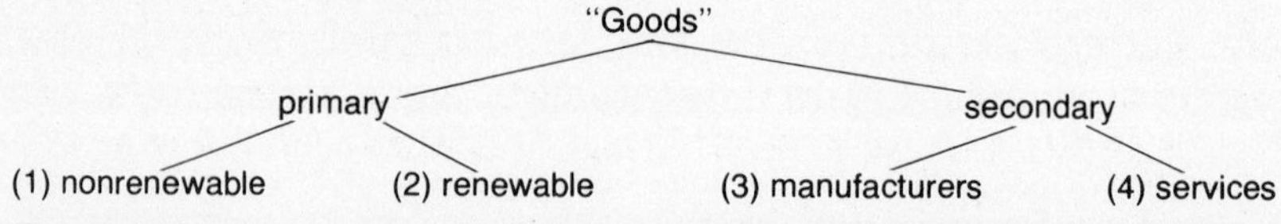

Economic theory is based on (3) and then carried over to (1), (2), and (4), rather than separate theories being developed for each. Another questionable set of economic assumptions, especially when applied to agriculture, are those related to the "trickle-down" theory.

the growing season in the wheat belt in Canada and the United States, then appropriate decisions to develop quicker-growing varieties have to be taken well in advance by crop breeders and seed corporations. By depending upon inappropriate economic models for our understanding of agriculture, we risk blinding ourselves to the many adaptive changes that are possible, and we encourage the even greater danger that our increasingly simplified and centralized agricultural system will collapse.

There are two obvious prongs to any strategy to reduce the risks inherent in the above situation (which obtains in other significant sectors as well). One is to develop new modes of thinking and analysis that better incorporate the real social, political, and environmental dimensions of agriculture and other policy arenas. Hopefully, the approach used throughout the book—contextual analysis—adds several useful dimensions to the debates and reassessment currently going on. As far as policy measures to decentralize our simple systems, it must be kept in mind that decentralization has to be accompanied by diversification, because it is the diversity of systems that gives them stability. Any policy to encourage a better balance between the urban and rural sectors has to go far beyond a simple physical dispersal. Ultimately this means the development of a number of basically different and locally adapted energy systems, transportation systems, industries, and farming systems. In each of these areas, there will have to be a reconsideration of the simple assumption that "bigger is better" and an attempt to design new technologies to fit the appropriate scale of operation.

The associations between technologies, their scale, their links with power, and their impact have been analyzed by E. F. Schumacher, who has suggested the name "intermediate technology" for the new kinds of technologies that are needed in each sector:

> As Gandhi said, the poor of the world cannot be helped by mass production, only by production by the masses. . . . The technology of *mass production* is inherently violent, ecologically damaging, self-defeating in terms of non-renewable resources, and stultifying for the human person. The technology of *production by the masses,* making use of the best of modern knowledge and experience, is conducive to decentralization, compatible with the laws of ecology, gentle in its use of scarce resources, and designed to serve the human person instead of making him the servant of machines. I have named it *intermediate technology* to signify that it is

> vastly superior to the primitive technology of bygone ages but at the same time much simpler, cheaper, and freer than the super-technology of the rich.[58]

To overcome the mutually reinforcing tendencies toward centralization and simplification that currently exist, what is needed are a series of policies—which we may call *framework policies*—that encourage decentralization and diversification. The difficulties involved in instituting and maintaining such policies can be seen in the antitrust field—where the basic pieces of legislation, the Sherman and Clayton Acts, are essentially framework policies designed to keep the marketplace decentralized and diverse so that the "natural" regulatory operation of the market will then, in fact, function. Without these precautions, the "hidden hand" of the market becomes that of a puppet—something manipulated by the string pullers (something that in fact has largely occurred).

There are, of course, a number of trends challenging the momentum toward centralization and simplification. The oil embargo, for one, has probably done more to force a wide-ranging reconsideration than any other single item. However, as suggested throughout this book, reassessment or establishment of a framework policy in one sector alone is not sufficient—witness antitrust policy. It is specialization itself—whether conceptual or infrastructural—that needs to be challenged, reevaluated, and placed within a larger framework. While it is easy to call for holistic approaches and for interdisciplinary work, these are of limited use as long as the basic structures of society are themselves highly specialized. Again, questions of size or scale are critical. At a global level, one can be holistic only in terms of very broad concepts, such as the hydrologic cycle. And at that level, one can employ only equally broad units of analysis that foreclose the capture of any great amount of detail. However, by employing simultaneously several different time-frames (and their concomitant concepts and units of analysis), contextual analysis permits one ultimately to trace the detailed interactions of a specific context as they fit into progressively more general space and time frameworks.

Even assuming, however, that contextual analysis were widely

[58]Ibid., p. 145. See also the valuable work by Nicolas Jéquier, ed., *Appropriate Technology: Problems and Promises* (Paris: OECD Development Centre, 1976). *Appropriate technology* now seems to be the most commonly accepted term.

adopted, and that the result was a much better representation of the various realities our lives intersect, this would still leave the basic dilemma of how to move to social, political, economic, and environmental systems that are more stable and offer the possibility of actually living more holistic lives. In short, contextual analysis appears to clarify greatly the risks of our dependence on huge but simple centralized systems that are highly unstable (and thus also highly unpredictable); it also appears to offer ways for conceptualizing alternative strategies for dealing with these risks. The ultimate tests, however, remain political—for people must be educated, persuaded, and pressured to accept those changes in their life-styles, work places, habits, and thinking that will move industrial society toward decentralization and greater diversity. The alternative would clearly appear to be some sort of major but largely unanticipated collapse, perhaps preceded by a period of increasing authoritarianism.[59]

International Implications

A final question remains in regard to the shifts that have been suggested for the industrial countries: How would such shifts affect the nature of the international system? Obviously, no precise answer can be given; however, some of the broad structural implications are fairly clear. To trace these, one must stress that the shifts and changes suggested above apply primarily to the industrial countries. Quite different shifts will be necessary in other regions, and the mix will vary between regions, not only because of their different historical contexts but because major changes in social structures and values feed through different cultural systems at different rates. Equally, major environmental changes will have quite different impacts on each region. In any case, it can be suggested that the kind of international system resulting from implementation of the shifts suggested above can be expected to be one much less dominated by the industrial countries. Current international security and economic structures are dependent upon highly centralized and sophisticated technologies and institutions.

[59]Robert L. Heilbroner, in his book, *An Inquiry into the Human Prospect* (New York: Norton, 1974), argues that a lack of social and institutional adaptability can be expected to lead to a period of strong authoritarianism. This, he seems to feel, might prevent a major collapse.

One thinks particularly of weapons-alliance systems and of multinational corporations–international monetary systems. The latter two increasingly will be brought into question as their social, energy, and environmental costs are clarified. The former will be reevaluated and downgraded as concepts of "national security" are expanded to include cultural, health, and environmental threats to national well-being.[60]

It should be noted that the suggested strategy of change is one of basic changes within states—especially the large industrial states—which then can be expected to "spill over" and ripple through the international system. This approach contrasts sharply with that of most futurologists, who simply assume a continuation of the centralizing tendencies of the past. What may be expected is a system much more open to genuine regionalism (i.e., regionalism based upon significant commonalities within the region, whether these happen to be cultural, economic, or environmental). In some ways, the process may parallel what appears to be happening in the United Kingdom, where after several centuries of centralization, there now is an attempt to devolve government and grant some regional autonomy to the Scots and the Welsh. While such a process at the global level is not incompatible with what Richard A. Falk has called "central guidance," it suggests that the content of such central guidance in the coming century should be more one of changing the nature of the constituent state actors than one of trying to manage their behavior structurally or physically from the outside.[61] The latter, if possible at all for any length of time, would appear to require rigid, highly centralized global systems that would be inherently unstable.

In short, we can conclude that so long as the major international actors (the industrial states) continue to centralize and simplify both themselves and the larger international system, it will be very difficult to make any predictions regarding the future—other than ones suggesting a variety of possible collapses. To the degree that matters of

[60]Lester Brown, *Redefining National Security* (Washington, D.C.: Worldwatch Paper No. 14, 1977).

[61]Richard A. Falk, *A Study of Future Worlds* (New York: Free Press, 1974), pp. 50–52. While the time horizon of the book is considerably shorter (1990) than the century of the developmental time-frame, the book offers a detailed and stimulating analysis of alternative world order structures (essentially institutional). It is also valuable in tracing interactions between preferred values and different modes of analysis.

scale and social and environmental costs can actually be internalized in major states and within their major industrial sectors, there is the possibility of establishing frameworks that will encourage diversity and a shifting away from our current overdependency upon energy-intensive technologies, centralized institutions, urban conglomerations, and specialized practices and thinking. As one reduces the scale of operations and the scale of interdependencies, the world becomes more predictable in the sense that chains of causation are shortened. There is a paradoxical aspect to this, for while one can say that movement in these directions will make local contexts much more predictable, one cannot predict in just which ways. This is due partially to the great variety of the real world and partially to the methodological limits of the developmental time-frame, where, as has been pointed out, it is primarily structural parameters and limitations that are discernible. In order to get a better assessment of the dynamics of specific contexts, we need to turn to the policy time-frame and an analysis of agricultural alternatives.

6

Agricultural Strategies and Policies for the Future

The various transformations of industrial society that are foreseeable and desirable over the coming century have been sketched out in the previous chapter. The evolutionary parameters into which those developmental changes must fit have been outlined. We now face the task of mapping out agricultural strategies and policies that are consistent with the limitations and goals of the two larger contexts. We must also give much greater attention to variations between different regions of the world than in the previous discussions. It should be noted that the mode of analysis employed here for deriving appropriate agricultural policy would be equally applicable to other economic sectors. The crucial point regarding the policy time-frame is that the kind of sectoral specialization that is possible at this level of analysis makes sense only when placed in the larger contexts of the developmental and evolutionary time-frames. Only by consciously assessing the larger societal and environmental trends, limitations, and interactions visible there can one hope to avoid the sort of sectoral tunnel vision that afflicts so much current policymaking.

STRATEGIC CONSIDERATIONS

In a sense, what is involved here is something analogous to outlining an "international agricultural decade." Essentially, the priority goals, the necessary data gathering, and the new policies needed all have to be developed, discussed, and then blended. The overall goal for agriculture is clearly the development of locally adapted agricultural production that is both sustainable and fruitful. This goal is consistent both with the genetic and environmental parameters of the evolutionary time-frame and with the need for less centralized and less energy-intensive modes of production foreseen in the developmental time-frame. It also is consistent with the sort of local and regional diversity that was shown to be both socially and environmentally valuable. In this regard, it should be pointed out that most of the discussion and analysis in this chapter will focus upon policies appropriate for the poor countries. This approach has a double use: first, it provides a useful contrast to the usual tendency (described in Chapter 3) to simply project Western agricultural conceptions and technologies upon the rest of the world, and second, it may offer some useful ideas for assisting in the coming transformation of industrial agriculture.

The Need for Alternative Data Systems

The first strategic aspect involved in devising locally adapted agricultural systems for the poor countries of the world concerns the question of relevant data: what data to gather, how best to gather them, and how to synthesize data from various specialties. Before examining these questions, it needs to be pointed out that there is an important political dimension integral to these data problems. That is, given the fact that information is power, how can the gathering and organization of data be structured so that it does not simply reinforce current maldistributions in agriculture but rather complements the goal of diversification and local adaptation? The tendency for Western natural and social scientists to see data as "neutral" and abstract makes it difficult to face up to this question. In addition, any serious approach to the question must necessarily involve the building up of new institutions, networks, and skills that will give the poor countries an independent capacity to assess and to influence the use to which such data are put. Such a data capacity must be independent in two ways: independent of dominant Western industrial conceptions and interests, and independent of dominant local elites.[1]

Keeping in mind these associated political problems, what are the appropriate approaches to selecting, gathering, and synthesizing data for locally adapted agriculture? First, it must be stressed that there is a close relationship between the purposes for which data are being collected and the scale of the data operations.[2] Evolutionary problems and processes require global concepts, categories, and data. Developmental problems and processes use less comprehensive concepts, categories, and data. Existing maps, surveys, and data collections include a mixture of the above plus others reflecting various policy concerns.[3] There is a great variation in coverage, reflecting past priorities: mineral and energy resources are fairly well known in terms of local, detailed surveys, while few countries have detailed maps or surveys of climate, water, annual precipitation, soils, vegetation, and animals.[4] The data needs for locally adapted agricultural production relate not only to scale (detailed and locally specific) but to its generally decentralized nature. This means that such data must be locally understandable (not too sophisticated) and adaptable (not so rigid and centralized as to exclude local variation). What is really suggested here is

[1]The United Nations *World Plan of Action for the Application of Science and Technology to Development* (New York: United Nations Dept. of Economic and Social Affairs, 1971) is useful for pointing out the current dependence of the poor countries upon the rich industrial countries for research and development of new technologies. However, even though it stresses the need for "appropriate technologies," the plan seems to assume that Western-style institutions staffed by indigenous scientists offer the best hope of developing locally adapted technologies. There is no discussion of the need for these new institutions to have some independence via-à-vis the local power structures and elites.

[2]United Nations, *Natural Resources of Developing Countries: Investigation, Development, and Rational Utilization* (New York: United Nations Dept. of Economic and Social Affairs, 1970), p. 11.

[3]See ibid. for a discussion of the various inventories and surveys being carried out by the UN family. Since the above report was written (1970), there have been a number of new systems added to monitor environmental data and climatic trends—primarily at the global and regional levels.

[4]Ibid., pp. 10–15.

the need for "intermediate" or "alternative" data systems and categories.[5]

There are many tensions and incompatibilities between this need and the general trend in data collection. Many of the technologies used for gathering, mapping, and interpreting data at the global and regional level are very sophisticated and carry with them the risk that industrial countries will control them and use them to gain political and economic leverage. This is especially the case with remote sensing satellites, where there is the additional risk that multinational corporations will be able to use their results more effectively—if they are publicly available—than most of the countries where the resources are located. Where this is the case, the companies have an immense advantage in securing leases for valuable resources. Also, even if ways can be found to assist the poor countries in doing general surveys of their basic resources, there remain problems of trying to prevent various national elites from exploiting them for their private benefit.

The search, then, is for intermediate data systems that will not be too sophisticated, too centralized, or too rigid—and that will draw upon local knowledge and priorities while synthesizing materials from the physical and social systems. While too expensive and too sophisticated in its current form, the approach of landscape architect cum ecologist Ian McHarg would appear to come the closest to meeting these various needs. In his *Design with Nature,* he shows how a series of shaded or colored overlays can be developed, which when combined show the most and least appropriate places for various types of land-use activities.[6] In his study on Staten Island, 30 overlays were prepared. These were grouped into categories of climate, geology, physiography, hydrology, soils, vegetation, wildlife habitats, and land use. Each of the 30 factors was evaluated on a scale ranging from low to high in presence and/or value. Each overlay was then shaded accordingly, with highest values darkest. By making composites from the overlays, a synthesis is achieved that shows the complexity of patterns and the areas of complementary or conflicting land use.[7]

[5]This assertion closely parallels the suggestion that "intermediate" or "appropriate" policies and institutions are needed to effectively promote intermediate technologies. See Nicolas Jéquier, ed., *Appropriate Technologies: Problems and Promises* (Paris: OEDC Development Centre, 1975), Chapter 6.

[6]Ian L. McHarg, *Design with Nature* (Garden City, N.Y.: Doubleday/Natural History Press, 1969).

[7]Ibid., pp. 103–115.

The advantages of McHarg's approach and methods are several. They enable one to draw upon a wide range of hard data (when available), and they indicate lacunae. They map the real spatial distribution of the various physical factors and permit their synthesis. Also, of great importance, they move away from simple economic measures, get at interacting natural processes, and encourage an incorporation of community values into the planning process:

> This method permits a most important improvement in planning method—that is, that the community can employ its own value system. Those areas, places, buildings or spaces that it cherishes can be so identified and incorporated into the value system of the method. Today many planning processes, notably highway planning, are unable to incorporate the value system of the community to be transected. At best, the planner supplies his own distant judgment.[8]

Finally, it is an approach that can be applied to different types of projects and to different scales (small cities to entire river basins). The net result is a system that if sensitively employed will capture complex processual interactions and that can be used to incorporate community values and local (and normally location-specific) knowledge about various aspects of the environment and the community. In McHarg's conception, the planner should not be the "outside expert" who arrives to impose a standardized "answer" (with all its attendant hidden values) but much more an entrepreneur who, by using this approach, is able to organize data and experts on location-specific processes and interactions and then encourage the local community to discuss and sort out its priorities regarding the area in question.

Choosing Appropriate Scales of Production

A second strategic consideration emerges after the data questions are dealt with: How do we determine the appropriate scales of production for locally adapted agriculture? Even after the basic climatic, geologic, hydrologic, and pedologic features have been surveyed and mapped out, one must evaluate different potential scales of production in terms of yield, efficiency, social costs and benefits, nutritional costs and benefits, and environmental costs and benefits. As suggested in Chapter 5, it is particularly misleading to evaluate agricultural scales of production by using yardsticks derived from industrial experience. For

[8]Ibid., p. 105.

example, one will come up with quite different estimates of appropriate scale depending upon the meaning given to only one of the variables: *yield*. If one is concerned primarily with monetary yield (the basis of most analyses), then a slight case can be made for larger-scale units—at least in the temperate zones.[9] However, if one is primarily concerned with maximum crop yield, then there is significant evidence to suggest that small, intensively cultivated units are the most productive.[10] Similarly, different scales will be suggested depending upon the meaning given to *efficiency*. In the United States, geographic and historical factors combined to lead farmers to seek to maximize return on labor since land was abundant. Over the decades, the search for labor efficiency led to more and more mechanization and chemical inputs. However, if a country's scarcest resource is land, then efficient use of the land becomes primary and may lead to labor-intensive farming on small plots. If water efficiency is the prime consideration, then other scales and systems are suggested. The latest limiting factor to be suggested is energy, and it is reasonably clear that the industrial countries will have to rethink their agriculture to try to make it more energy efficient.

There are also a number of important social and political dimensions that have a bearing on the choice of appropriate production scales. While some might question the utility of discussing the "choice" of production scales—given the general reluctance of governments to engage in land or tenure reforms—the matter is much more than academic. First, tax policies, credit arrangements, irrigation schemes, agricultural subsidies, etc., all have a long-term impact on farm size, and if the reigning assumptions are that scales of production are neutral, then larger farms will tend to be favored.[11] Second, given the links between land and social and political power—especially in

[9]It is not clear whether this conventional wisdom will withstand the feed-through impacts of higher energy costs.

[10]For evidence on the developing countries, see Keith Griffin, *The Green Revolution: An Economic Analysis* (Geneva: United Nations Research Institute for Social Development, Report No. 72.6, 1972), pp. 31–40. Among developed countries, the yield per hectare of small-scale, labor-intensive farms is more than twice that of the large, land-intensive farms. See Janos Horvath, "Food Production and Farm Size: A Reconsideration of Alternatives," Holcomb Research Institute Working Paper (Indianapolis: Butler University, 1977).

[11]See Griffin for a critique of such assumptions.

the developing countries—one cannot talk about changing the size of farms without raising basic social and political questions. While the social and political costs associated with the green revolution have received only passing comment by economists, there is a growing body of literature documenting the social disruptions and the increasing economic inequalities that have resulted from it.[12] As with the environmental costs of agriculture, these have been deemed "externalities" in terms of economic theory and thus do not appear in the cost/benefit analyses of corporations and economic planners. There are several levels of cost. One of the most serious costs associated with the green revolution—one that is only now receiving some official attention—is the loss of human capacity resulting from malnutrition. Most striking are the high infant mortality rates in many poor countries. Equally distressing is the impairment of the mental capacity of infants resulting from protein deficiencies while their mothers are pregnant or during the first six months of life.[13] The most direct link between malnutrition and the green revolution lies in the loss of pulses and other protein-rich crops as they are displaced by HYVs. This loss combined with the monetization of agriculture often means that the poor no longer have available to them locally produced and barterable proteins, nor can they afford commercial sources of protein. Complicating this outcome even more are general trends toward industrialization, urbanization, and new standards of "modernity"—such as the giving up of breast feeding. The need for the incorporation of nutrition in development planning is clear, if difficult and complex.[14]

Environmental considerations that relate to the appropriate scales for production are equally complex. What must be evaluated here is the "fitness" of the man-made to the natural environment. One needs

[12]See the two reports, based on a number of country studies, done by the United Nations Research Institute for Social Development (UNRISD) on *The Social and Economic Implications of Large-Scale Introduction of New Varieties of Foodgrain: Summary of Conclusions of a Global Research Project* (Geneva: UNRISD, Report No. 74.1, 1974); and Andrew Pearse, *An Overview Report* (Geneva: UNRISD, Report No. 75/C.11, 1975). See also, Francine Frankel, *India's Green Revolution: Economic Gains and Political Costs* (Princeton, N.J.: Princeton University Press, 1971).

[13]See Alan J. Berg, *The Nutrition Factor: Its Role in National Development* (Washington, D.C.: Brookings, 1973), and the articles on nutrition in the special food issue of *Science* 188(9 May 1975).

[14]See Johanna T. Dwyer and Jean Mayer, "Beyond Economics and Nutrition: The Complex Basis of Food Policy," *Science* 188(9 May 1975):566–570.

to know not only the basic parameters of the natural environment but the ways in which man-made arrangements encourage or discourage good yields and sustainable production. One basic dimension here is the complexity of the landscape itself as well as that of the cropping systems. For example, in England, there is concern that the simplification of the landscape and the reduction of hedgerows that result from the expansion of industrial agriculture and monocultures is leading not only to an aesthetic loss but to agricultural losses as well.[15] A similar environmental concern in many developing countries relates to the soil and water runoff losses that result from overgrazing of pastures and, in some cases, the stripping of trees. The importance of soil and water systems has already been stressed (Chapter 5). Shorter-term manifestations of improper approaches can be found in the loss of vital trace minerals in soils, the degradation of humus, the overcompaction of soils, declining water tables, water and soils made saline, and the eutrophication of lakes and reservoirs. A great many of these environmental problems are related to or shown to be exacerbated by large-scale, mechanized agriculture. Obviously, they can occur from overuse of traditional methods of agriculture as well, but generally they result more from population pressures than from inappropriate technologies.

Centralization versus Diversity

A third strategic consideration that permeates any attempt to formulate locally adapted agricultural policies is the question of the relative advantages of centralized as compared to decentralized and diverse approaches. This question is the organizational analogue of the functional concerns discussed in regard to scales of production. Variations in the degree of centralization and diversity are to be expected (and encouraged) between different regions of the world and between different sectors within countries. As has always been recognized in theories on federalism, there are some things necessarily centralized in

[15]See Jon Tinker, "The End of the English Landscape," *New Scientist* 64(5 December 1974):722–727, which discusses the Countryside Commission's report on "New Agricultural Landscapes." The report documents the continuing rapid changes in the countryside—especially the loss of hedgerows—due to the pressures of industrial agriculture.

a governmental system, while there are many others that are best decentralized. The real art of federalism lies in sorting out which is which according to time and place; thus, for example, Tocqueville stressed that because of its history, location, and neighbors, France would have to be more centralized than the United States, even though he clearly recognized the risks of centralization in democratic societies.[16]

In regard to the poor countries, with which we are primarily concerned here, there exists a very curious mixture of centralization and diversity. The colonial heritage of most of them has left a thin layer of centralized modern institutions and infrastructure that have been used by the elites to promote "nation building." The diversity and decentralized nature of almost all other aspects of these societies is often overlooked or underestimated by Western development planners, who generally see them as an obstacle to modernization. It is curious but very common that people seeking to draw upon the experience of others or to transport their own experience tend to focus on the results or the end product. For example, many Latin American countries adopted the U.S. Constitution almost verbatim in the 19th century, and today attempts are made to export the institutions of the European Communities ("Common Market"). What really needs to be examined and drawn upon in such cases is the *process* by which results were achieved. Thus, it would seem clear that since the process of development in the West was one of progressively building industry upon agricultural advances, and more centralized institutions from confederations of well-organized local units, the primary focus of development should be upon grass-roots agricultural activity and organization. While such a task is very difficult, it offers better long-range prospects than attempts to bypass basic agricultural and rural development.

The appropriate strategic mix of centralization and diversity for each sector will vary from country to country; however, in line with the developmental parameters stressed in Chapter 5, some general points can be made. In the political sector, it would appear that the most neglected level is that of local government. In the poor countries,

[16]Alexis de Tocqueville, *Democracy in America* (New York: Vintage, 1945), Vol. 2, pp. 334–339.

everything conspires against the giving of any significant role to local government. Bureaucratic structures carried over from colonial times still are designed primarily to "preserve order" above all else. The attitudes of those manning these structures often reflect old colonial notions of administrative order plus a new admixture of technocratic superiority. The bureaucrat and the planner know what is best for the peasant and how he is to be fitted into various grandiose development plans. In most cases, there is both governmental and administrative centralization, a mixture whose negative effects Tocqueville remarked upon in regard to France.[17] As will be seen later, many of the agricultural projects that have been most successful have involved getting the peasants or their self-selected representatives to become their own agents in dealing with bureaus and officials, rather than simply being the final recipient of orders flowing down a long chain of command. Such a shift on the part of peasants is, of course, potentially threatening to local landlords, moneylenders, and other vested interests—which helps explain the reluctance of national governments (often dependent upon these local groups) to encourage changes in this direction.[18] Equally, regional or international organizations are especially cautious about interfering in the "domestic affairs" of member states. Thus, to the limited degree that developmental or agricultural planning directly seeks political change, it is through the subterfuge of hoping that infrastructural or economic policy changes will encourage appropriate political responses—a hope that is usually vitiated because of false assumptions regarding the neutrality of market mechanisms and scales of production. This situation suggests two things for regional and international organizations: first, that they must formally and more forcefully incorporate criteria of political change and participation in their programs, and second, that they need to become much more sophisticated regarding the real redistributive effects of various kinds of policies and technologies.[19]

The situation in the economic and educational sectors is similar—although another element soon appears that is closely linked with any

[17]Ibid., vol. 1, pp. 89–101.

[18]The parallels with the War on Poverty requirements for inclusion of the poor in program planning are clear. The reactions of local political bosses were strongly negative—as was to be expected.

[19]Unfortunately there is not a great deal of material available upon which they can draw—a commentary in itself on the academic world.

attempt to encourage diversity and decentralization: the spatial distribution of installations and institutions. Only recently have the spatial aspects of development received much attention.[20] Most of the work done in this area has involved mapping out the correlations between industrial development and the spatial distribution of infrastructure. Much less has been done in examining the spatial distribution of schools or in relating infrastructural development to agriculture, especially subsistence agriculture.[21] Those studies that have been done indicate the great importance of "establishing a large number of small growth centres as foci of innovation, rather than a few large and relatively isolated improvement schemes."[22] The difficulty here—as with all the other sectors—is persuading capable people and officials to work on an extended basis in the rural areas. The attractions of an urban location in terms of salary, status, and political influence reinforce many of the antirural attitudes acquired in the course of a westernized education. What is needed is a strategy that will ultimately make rural living much more attractive. In the earlier stages, this may well mean personnel policies designed to encourage work in the boondocks, although educational reforms and the generation of rural economic opportunities may be more important in the longer term.

Another spatial concern—though it is not usually thought of in those terms—is that of land use and land reform. Although land-use laws and land reform involve a number of economic and social dimensions, ultimately they are concerned with who controls the use and product of specific parcels of land. Land-use decisions and powers are normally fragmented among a wide range of agencies (zoning, public health, road construction, agriculture, forestry, etc.); likewise the legal, administrative, and political procedures for resolving land-use conflicts are complex and vary from issue to issue, as well as from country to country. Land reform, while less broad than many land-use questions, is more fundamental since "in the traditional and generally accepted sense of the term, land reform means the redistribution of

[20]Brian S. Hoyle, ed., *Spatial Aspects of Development* (New York: Wiley, 1974).

[21]See Arthur T. Mosher, *Creating a Progressive Rural Structure* (New York: Agricultural Development Council, 1969), for a detailed discussion of the importance of spatial and infrastructural factors. However, his goals and strategies are different than those presented here since he is promoting modern industrial agriculture, where the farmer is progressively integrated into a national economy.

[22]Ibid., p. 16.

property or rights in land for the benefit of small farmers and agricultural labourers."[23] The need for land reform is often mentioned in agricultural studies, but normally only in passing and more in terms of pious wishes than in concerted, detailed planning. While the political obstacles to land reform may explain some of the reluctance of policy makers to consider it seriously, this is not a sufficient explanation, since in many other politically sensitive areas, policy makers and bureaucrats have no such hesitations. While there are ideological hesitations in many Western-orientated countries, one of the most important reasons for the lack of proper consideration of land reform lies in the fact that it is basically an institutional factor and thus tends to be ignored by conventional economists. In his study demonstrating that land reform generally aided development in Taiwan, Anthony Y. C. Koo points out the rareness of studies examining the impact of land reform upon development:

> One reason for the lack of such study is perhaps that land reform takes the nature of an innovation in the social order. It changes the organizational and structural framework within which economic activities occur, whereas existing economic theories usually assume that the institutional framework is to be held stable in order that analyses can be reduced to manageable proportions.[24]

Even so, land reform should be a primary concern of policy makers. While fundamental changes often occur only in times of great turmoil or crisis or as a result of occupation by foreign powers, consistent policy measures over a period of years can shift the emphasis measurably.[25] Thus, policy makers should have the general directions in which they want to move in land reform well worked out so that they can seek incremental change and, should the opportunity present itself, more fundamental change.

In any case, the ultimate success of any rural development program rests upon the degree to which the millions of peasant farmers can improve their own lives, farms, and production. Just as the farmer

[23]Doreen Warriner, *Land Reform in Principle and Practice* (Oxford: Clarendon, 1969), p. xiv.

[24]Anthony Y. C. Koo, *The Role of Land Reform in Economic Development: A Case Study of Taiwan* (New York: Praeger, 1968), p. 2.

[25]For historical examples, see Warriner, and Ronald P. Dore, *Land Reform in Japan* (London: Oxford University Press, 1959), who discusses both the reforms of the interwar period and those of the MacArthur occupation.

must organize his life around the basic requirements of his crops—proper tillage, planting at the right time, weeding, harvesting at the right time, etc.—so too should strategic planning for the agricultural sector be organized around the basic needs of adaptive agriculture. This has rarely been done on a consistent basis; what it really requires is a shift from emphasis on trying to control or change the environment to emphasis on better adapting production to the local environment. Historically, this was the process involved in the temperate zone, whereby there was great experimentation with different crops, varieties of seed, cultivation techniques, rotation patterns, etc. In the United States, this was facilitated by the localized institutional support systems available to farmers through the land-grant colleges, research stations, and extension programs. Thus, the basic process was one of drawing upon traditional agricultural crops, seeds, and practices and gradually adapting them and improving them. Ironically, however, attempts to improve agriculture abroad have tended to fall into the trap of trying to export the result rather than the process. Not only have we attempted to export temperate zone crops and practices, but perhaps more seriously, our whole understanding of tropical agriculture has been filtered through quite opaque temperate lenses:

> Nearly all research in tropical agriculture is highly reductionist, parochial, and discipline-oriented. This can be quickly observed by perusal of [the basic texts in the field], as well as tropical agricultural journals. Articles with a holistic approach are a conspicuous rarity in the trade journals.[26]

Thus, there is a need for both a general reconceptualization and a whole range of new policies. The goal of locally adapted agriculture clearly suggests a series of basic shifts in emphasis: from large irrigated farms to small peasant farms (often dry-land); from large-scale industrial agriculture to small-scale diverse farming; from national development needs to local needs; and from the exportation of Western intellectual models and technologies to the development of local resources and skills.

Obviously, full backing for the above suggestions will be most difficult to obtain. At the national level, giving a much higher priority to agricultural development represents not only a major shift in

[26]Daniel H. Janzen, "Tropical Agroecosystems," *Science* 182(21 December 1973):1212. This article is an excellent survey of all of the various misunderstandings temperate zone observers have of tropical agriculture.

priorities but a threat to the position and status of many bureaucrats, planners, and politicians. Although there are analogous problems among international agencies and foreign aid programs, one can see a creative role for them if they can only disenthrall themselves from their past policies and priorities. In the key area of agricultural research, much can be done not only to build greater flexibility into the HYV package but to do research on small-farm systems in various climates and in various cultures. Rather than trying to find the one or two "key" elements (which it is then hoped can be transferred to increase production everywhere), the overriding conception should be one of establishing localized research and support infrastructures that will promote sustainable increases in the well-being of specific groups of peasant farmers. The shift from a quantitative measure of output as the prime concern to one of the qualitative well-being of the peasant and his land is fundamental. Such a shift brings into active consideration everything from nutrition to long-term environmental degradation.[27] While there is a range of possible policies and institutional formats that can be employed to pursue such goals, the main challenge at the international level will be to reconceptualize current agricultural assistance and research programs.

POLICY CONSIDERATIONS

As with the other time-frames, the ultimate goal in the policy time-frame is to develop an understanding of the specific context—the momentum of past trends plus the charting of future constraints. Thus, in working out actual policies, one expects to end up with quite different mixtures or packages of programs from region to region or country to country. While some examples of relatively successful packages will be given in the next section, it is necessary here to discuss some general policy considerations. Although these do not give any precise indication of how to put together an appropriate package, they do show the range of considerations that must be included. It should

[27]The summary of the project on the green revolution carried out by the United Nations Research Institute for Social Development illustrates the range of concerns such an approach involves. See UNRISD, *The Social and Economic Implications . . . Summary of Conclusions.*

be noted that the various policies discussed here deal primarily with the agricultural sector, although success in the agricultural sector obviously requires supporting or complementary policies in other sectors. Also, the discussion will emphasize those considerations that have received the least analysis in the conventional literature on agricultural development.

Conservation of Natural Resources

Turning first to policy issues related to natural systems, it is clear from the prior discussion of developmental limitations and parameters that all such policies have to be based on the conservation and efficient use of natural resources.[28] Also, at the policy level, one must be especially conscious of how resources are allocated. The large strategic task of obtaining data on the physical location, distribution, and extent of natural resources is assumed to be under way and the decision makers, as always, to be operating with less than complete data. Turning first to questions of water policy, one must assess what are the most appropriate systems for supply and distribution. In choosing the most appropriate supply systems, problems of drainage and salinization must be considered. Whether one should focus on irrigation or dry-land techniques is another question. The problem of large versus small systems requires one to consider alternative approaches to water management, whether traditional or modern. These include the harvesting of rainwater, various conservation techniques, trickle irrigation, and the use of underground reservoirs and irrigation systems.[29] Distribution questions relate both to the macrodistribution of water systems throughout a country and to the microdistribution of water within each system. The macroquestions really relate to regional problems (dry-land versus irrigated areas and how much to develop each) and to class problems (how much of an attempt there should be to reduce gaps in the standard of living between various groups). The microproblems involve timing of water distribution, ways to prevent

[28]For a general discussion of the need and utility of conservation, see *Natural Resources of Developing Countries,* pp. 37–42.

[29]For a valuable compilation of alternatives, see Board of Science and Technology for International Development, *More Water for Arid Lands: Promising Technologies and Research Opportunities* (Washington, D.C.: National Academy of Sciences, 1974).

abuses by large users, private versus public financing, competition between agriculture and industry, pollution questions, etc.[30]

These sorts of policy questions make clear the fact that in developing water policy, there are a number of other considerations that have to be included. The tendency in water policy—as in energy policy—has been to take the basic resource itself too much for granted and to let other policy matters predominate. Particularly as the price of energy goes up, so too will the price of water—both coming somewhat more in line with their real value. These price questions also tie in with questions of interdependence. Does a country want to tie its water policies to the availability of oil? To the availability of sophisticated foreign technologies such as nuclear desalinization plants? What are the costs of trying to develop indigenous water systems with minimal external influence? Also, from a regional or global perspective, serious questions have to be asked about large-scale dams. Not only do they potentially foster regional climate shifts (see Chapter 5), but they are much less efficient in arid or semiarid zones, where up to 50 times as much water may be needed for a given crop as in the temperate zone. This suggests that it would be better to use scarce international resources to improve dry-land crops, to promote alternatives to conventional irrigation, to improve trade between regions, etc. It would appear that policies aimed at rural development and at the well-being of the numerous small peasants would reduce some of the conflicts between national desires for independence and the comparative advantages and disadvantages of world trade. The poor countries are at a comparative disadvantage primarily when they try to compete with the rich in industrial agriculture. By focusing on the nonindustrial or peasant sector of agriculture for their major developmental thrust, the objectives of individual poor countries and the major international aid-givers are less in conflict.[31]

[30]For a valuable discussion of how water policy (as well as such other agricultural services as credit, extension, and marketing) need to be reoriented if both conservation of resources and greater social justice are to be achieved, see *The Social and Economic Implications . . . Summary of Conclusion*, pp. 32–46. In regard to water policy, it is suggested that investment should be in smaller-scale systems and in improving existing systems, especially field distribution systems. Labor-intensive methods of doing this are recommended, as are various mechanisms (such as graduated rates for irrigation water and acreage ceilings) that would be aimed at encouraging the small rather than the large cultivator.

[31]This would be the case at the level of formal policy statements. It would still be very difficult to wean the international agencies away from the idea of integrating member

Many of the same policy considerations come into play in the energy sector as in water systems, although the abundance of petroleum in the OPEC countries stands in sharp contrast to the effective availability of water in most semitropical regions. The costs of energy interdependence to those farmers in the developing countries who have adopted industrial modes of agriculture have become painfully clear. Fuel for tubewell motors has become very expensive and scarce at the same time that fertilizer prices have increased dramatically. Not only are there fluctuations in the price of major crops according to supply and demand, but today even fluctuations in currency exchange rates can have a great influence on the financial success or failure of a farmer. As indicated earlier, serious consideration of the relative energy efficiency of different modes of agriculture has begun only in the last few years. Early indications are that traditional modes of agriculture are more energy efficient and can be more productive as well (in terms of crop yield per acre). This finding suggests that policy makers (both national and international) should encourage improvement of traditional modes. It also suggests that the typical Western approach of promoting the rapid electrification of the rural areas needs to be reexamined in light of alternative sources of energy such as methane converters, windmills, and solar panels, which may not require such large infrastructural investments and may be more energy efficient.[32] In the energy area especially, policy makers should seek to keep as many options open as possible.

The importance of soil conservation has long been recognized at the national and the international level. Even so, the process of trying to implement effective soil conservation measures has been difficult. This is most obvious in the example of the U.S. Soil Conservation Service—one that is often cited favorably. The history of the SCS has been one of controversy, organizational turmoil and shifts, and conflicting objectives.[33] Starting out in the Department of the Interior, the

---

countries more and more into the global economy, just as it will be difficult to wean many industrial-oriented elites in the poor countries away from their comfortable links with the industrial countries.

[32]For general discussions of alternative energy sources, see Board of Science and Technology for International Development, *Energy for Rural Development* (Washington, D.C.: National Academy of Sciences, 1976), and Arjun Makhijani, *Energy and Agriculture in the Third World* (Cambridge, Mass.: Ballinger, 1975).

[33]Robert J. Morgan, *Governing Soil Conservation: Thirty Years of the New Decentralization* (Baltimore: Johns Hopkins Press for Resources for the Future, 1975), describes all of these in detail.

SCS was soon moved to the Department of Agriculture. Over the years, there have been conflicts within the Department of Agriculture regarding the role of the SCS, especially its relationship to various state programs and to the land-grant colleges, as well as to how its work could be coordinated with that of the Agricultural Extension Service.[34] Questions regarding whether it should be a specialized service or a multipurpose planning mechanism have inevitably become intertwined with local, state, and national politics. Also, many of the phenomena normally associated with new programs in the developing countries—the rich getting richer, programs being dominated by local elites, etc.—have been visible in the SCS.[35] While there have been generally favorable assessments of the overall environmental impact of the SCS programs, the available data make a wide range of interpretations possible.[36] Similar problems can be expected in any country that seriously attempts a decentralized soil conservation program. Particularly in the area of soil conservation, there is great logic in adopting a watershed approach once one goes beyond attempts to help local farmers improve their individual farms. The benefits of doing this even within a federal system are illustrated by the Tennessee Valley Authority. In regions of the world where watersheds cross several national boundaries, political problems can be expected to be difficult.[37] Some of the policy difficulties associated with soil conservation have been emphasized here because it is an area where there is a great deal of experience available. As in most areas of human experience, it is easier to document and comment upon the current difficulties than to assess fairly the successes and the disasters avoided. In any case, the record is encouraging enough to suggest a continued striving to improve soil conservation—as well as to expand our general concepts and organizational approaches to conservation.

[34]Ibid.

[35]Ibid., especially pp. 317–319 and 349–352.

[36]R. Burnell Held and Marion Clawson, *Soil Conservation in Perspective* (Baltimore: Johns Hopkins Press for Resources for the Future, 1965), pp. 320–321, state that "the available data permit an estimate that the soil conservation job as it appeared in the early thirties had been 20% completed by 1958 or had been 60% completed—depending upon which of equally reasonable assumptions are used."

[37]Although the Mekong River Project is often cited as evidence to the contrary, one can expect that social, economic, and environmental problems will become highly politicized should the project move from its current largely planning phase to any significant degree of implementation.

Genetic conservation is another area that has received increasing attention, at both the national and the international level. Efforts at such conservation have been largely divided between those concerned with wildlife and those concerned about plant species, especially crop plants. The greater ease with which wildlife conservation programs can gain public support is demonstrated by the increasing numbers of wilderness areas, game preserves, and bans on the sale of products from endangered species. Even though the threat to the genetic diversity of crop gene pools would appear to be greater and the impact upon human affairs more immediate, little has been done beyond the establishment of a number of gene banks. Gene banks face three major risks: human error, mechanical failure, and removal from normal selective pressures (see Chapter 3, pp. 80–81). Thus one observer has concluded that

> preservation on site is the best method and should be our first priority in international agricultural and environmental management. Since plant populations, not entire ecosystems, are being preserved, *in situ* preservation need not encompass large amounts of land. Carefully chosen strips of 5 by 20 kilometers (approximately three by twelve miles) at as few as 100 sites around the world where native agriculture would be continued might be adequate.[38]

Such a proposal may be viewed as the minimum required for conservation of basic food plant stocks. However, given the various evolutionary and developmental trends spelled out in Chapter 5, national and international planners should also think about policies that will encourage a decentralization of infrastructures and a diversification of agricultural modes. *It must be remembered that reduction in the number of ecosystems is as great a threat as genetic simplification (reduction of the number of crop species).*

Policy makers, especially in the developing countries, also need to be concerned about the institutional methods by which agricultural inputs and outputs are handled. The various difficulties and potential risks involved in seed multiplication and distribution, fertilizer production, fuel provision, equipment design and production, food processing and distribution systems, and international marketing ar-

[38]H. Garrison Wilkes and Susan Wilkes, "The Green Revolution," *Environment* 14(October 1972):39. In many ways, this suggestion would be similar to legislation being proposed in Switzerland to pay mountain farmers a subsidy to keep them on the land, practicing their soil-conserving agriculture.

rangements have been described in Chapter 4. While most of these involve social and economic policy more than policy for natural systems, the area of seed multiplication and distribution involves a key natural resource. Policy decisions in this area also can have a profound impact upon the relative dependence or independence of a country's agricultural policy, just as decisions in the social and economic areas can have a serious impact upon a country's ecosystems.

Discussion of policy considerations related to social systems will be brief. For one thing, many have already been touched upon. For another, there is an abundance of such discussions—although most of them are flawed in at least one of three ways. Typically, they are written without adequate awareness of how natural systems operate and interact with social systems. Rarely do they try to estimate in advance the distributional impact of innovations as they spread differentially in terms of space, time, and resource availability. Finally, as stressed throughout the book, there is little awareness of the cultural baggage associated with the technologies and policies that it is proposed be exported.[39] Many policy prescriptions also suffer because they are made in terms of "universal" functional categories that are only partially applicable and/or adaptable to specific environmental and cultural contexts.

How to Integrate Policy at the Local Level

As an introduction to the case studies to follow and to illustrate how one can integrate social and natural policy concerns, let us examine Egbert de Vries's concept of the "intensive development zone."[40] There are several key aspects to this approach. It focuses on a specific geographic region and assumes detailed knowledge of its environment. Organizationally, it tries to overcome bureaucratic fragmentation by setting up five or six broad functional agencies for the zone. Socially and economically, it seeks to make peasants more pro-

[39]Typical examples include Theodore W. Schultz, *Economic Growth and Agriculture* (New York: McGraw-Hill, 1968); Herman M. Southworth and Bruce F. Johnston, eds., *Agricultural Development and Economic Growth* (Ithaca, N.Y.: Cornell University Press, 1967); Max Millikan and David Hapgood, *No Easy Harvest* (Boston: Little, Brown, 1967); and Clifton R. Wharton, Jr., ed., *Subsistence Agriculture and Economic Development* (Chicago: Aldine, 1969).

[40]The following description is based on discussions with Professor de Vries.

ductive and self-sufficient. Underlying this goal is the historical recognition that "the poor are the victims of development" and that the overall policy package must ensure that benefits will actually accrue to the poorest segments of the population in the zone. Also, in contrast to many development theories, the economic goal is not one of integrating the peasants totally into market mechanisms but rather one of giving them some cash income beyond their basic agricultural self-sufficiency. The main incentive to the peasant is to provide an education for his children. A rough mix of approximately 60% nonmonetary income (basic self-sufficiency) and 40% monetary income is suggested, and this means that there must be a "key" crop to provide this tangible increase in cash income.

The overall policy package obviously must go beyond agricultural concerns, and in many ways, the approach parallels that of those advocating rural development, although there is a more consistent concern with integrating natural systems.[41] Three different cost-benefit calculations are involved: (1) conventional economic calculations; (2) assessment of labor requirements and impacts—including secondary and tertiary effects; and (3) evaluation of energy efficiencies and leakages. Educational and labor needs can be met in a variety of ways and must be tailored to the culture and environment. For example, Iceland has used tax policy to encourage children to work on family farms while earning money for educational purposes. At age 12, a child becomes a taxpayer and must also be paid the minimum wage; however, a farmer may then deduct the cost of his child's wages on his own income tax return. As a result, children earn school money while at the same time being encouraged to work on the family farm. Another approach aimed at discouraging rural outmigration is found in Indonesia. In the Karangmojo district of Central Java, a project has been set up to combat deforestation and soil erosion (as well as to replace weaving jobs lost to automation) by planting mulberry trees in cooperation with private Dutch and Canadian aid as well as Japanese foreign aid. The Japanese were to get part of the silk crop in return for their technical assistance, while the local weavers (usually women working in their villages) were to get the rest. Part of the project involved children of school age picking some mulberry leaves each

[41]See, for example, Edgar Owens and Robert Shaw, *Development Reconsidered* (Lexington, Mass.: Lexington Books, 1972).

day, which they then could exchange at school for their noon meal. The overall package thus encourages reforestation, local employment, and generation of foreign exchange and provides incentives for the education of local youths. The overall success of the project is perhaps best seen in some of the adaptations made to the basic plan by the villagers. As Professor de Vries reports:

> Last November [1976], I visited the area. . . . There have been non-governmental Dutch and Canadian contributions, Japanese technical assistance and World Food Programme aid. Interestingly, while "we" thought of the Mulberry as a source of silk, the villagers eat the fruits and the young leaves, broil the caterpillars as animal protein and boil the cocoons—eating the pupae, which taste like shrimp! They did better than we thought! I saw a class of third grade school children (the school was built with $1000 of Canadian funds) and all seemed well fed, healthy and happy—an enormous contrast with the starving youth of eight years ago![42]

In conclusion, one can note that Professor de Vries's localized approach is entirely compatible with intermediate technologies and generally fits the requirements of the contextual approach suggested here.

GOING BEYOND RURAL DEVELOPMENT

Rural development is a concept that has received a good deal of academic and governmental interest in the past few years. It represents a major advance over conventional development theories, particularly in its critique of them. The main theme of this critique has been that conventional concepts and programs greatly overestimated the importance of the industrial sector and that to the small degree that they dealt with rural development, they focused on the few irrigated farmers to the detriment of the rest.[43] Important as these points are,

[42]Communication from Egbert de Vries, dated 24 January 1977. The UNRISD report, *The Social and Economic Implications . . . Summary of Conclusions,* pp. 46–48, suggests an "area development unit" as a similar way to integrate local services in the public sector. Three major functions are identified: (1) an entrepreneurial complex that would assist the small cultivator in obtaining seeds, fertilizer, credit, crop storage, etc.; (2) a localized research-and-development unit that would carry out resource inventories, develop appropriate technologies, do field testing, advise on plant protection, etc.; and (3) shared capitalization for public-works-type projects, especially irrigation, to avoid their monopolization by large private interests.

[43]Three rather different views of rural development are found in Owens and Shaw; Laurence Hewes, *Rural Development: World Frontiers* (Ames: Iowa State University Press, 1974); and René Dumont with Marcel Mazoyer, *Socialisms and Development* (New York: Praeger, 1973).

they need to be supplemented in two important ways if we are to move toward locally adapted and sustainable agriculture. First, as indicated above, new ways must be found to survey and synthesize data regarding the environmental, cultural, and social aspects of rural life so that policy will be better based. Second, there is a need to reexamine several historical examples of fairly successful rural development in light of the above concerns to see what additional lessons for policy makers can be gained.

Small-Scale Attempts: The Comilla and Puebla Projects

In the brief case studies that follow, examples of small-, medium-, and large-scale programs will be included, thereby clearly illustrating the various ways in which policy problems vary according to the scale of the operation. Throughout, well-known examples have been chosen to minimize the need for detailed description and to increase the analytic discussion. For the small-scale projects, those in Comilla, Bangladesh, and Puebla, Mexico have been chosen. The Comilla project is located in a typically overcrowded, poor region of Bangladesh, which at the outset of the project in 1961 had a population of some 217,000, including some 56,000 farm families occupying approximately 52,000 acres of farmland.[44] The project was initiated by a prominent Pakistani civil servant, Akhter Hameed Khan, with the establishment of the Academy for Rural Development. Khan's goal was to utilize the social sciences to promote rural development. His approach was one of both institutional transformation and institutional creation. On the govenment side, he sought to transform the existing agencies at the county level from aloof, specialized, and largely urban-oriented fiefdoms into an integrated delivery team.[45] To accomplish this, the head-

[44]Most of the details on the Comilla project are drawn from Robert D. Stevens, "Rural Development Programs for Adaptation from Comilla, Bangladesh," *Agricultural Economics* (East Lansing: Michigan State University, Department of Agricultural Economics, Report No. 215. June 1972); and Millikan and Hapgood, *No Easy Harvest*. For a detailed history see Arthur F. Raper, *Rural Development in Action* (Ithaca, N.Y.: Cornell University Press, 1970).

[45]Stevens, p. 7, gives the following characterization of the civil service: "The 'iron framework' of Pakistan had generally been trained to focus on the limited problems of law, order, and taxation. There is consensus that the high-ranking officers of this civil service were highly intelligent and well educated individuals. Most of them, however, had urban biases and little knowledge of rural development needs and programs."

quarters of the various county administrative and technical services were physically moved to the academy.[46] The various bureaucrats thus had daily contact not only with each other but also with the staff of the institute. In addition, the academy carried out teaching and research programs that were designed to give in-service training to higher-level bureaucrats. To facilitate this, the board of governors of the institute was structured to include government-level bureaucrats from each of the major users of the academy's services.[47]

On the peasants' side, there was a concerted attempt to create new institutions that would be more decentralized and genuinely representative of the peasants' interests than the old extension services and cooperatives had been.[48] Behind these shifts lay a combination of practical and political motives:

> In our opinion the cooperative was the best vehicle for extension as well as for supplies and services. At the same time we thought, rather behind the back of our government, that these cooperatives should also protect the peasant members from the prevailing system of money lending and trading. It did not seem feasible to do anything, at this stage, about lease rates or other tenure matters.[49]

The new cooperatives were critical in gaining the support of the peasants and for increasing the communication between the "two cultures" that one finds in most developing countries, the urban-literate culture and the rural-oral culture. Beyond these changes in organizational infrastructure (something that rarely seems to get attention comparable to that given physical infrastructural needs), the integrated and comprehensive nature of the programs developed by the academy was

[46]In fact, even residential quarters for county officials were provided at the center. Akhter Hameed Kahn, *Reflections on the Comilla Rural Development Projects* (Washington, D.C.: American Council on Education, Overseas Liaison Committee Paper No. 3, March 1974), pp. 24–25.

[47]Ibid., p. 5.

[48]The old cooperatives included 10–15 villages and had a government-appointed manager, while the extension service depended upon a government employee living in the village as the point of contact with farmers. The new cooperative system was organized with the village as the basic unit, where the members (typically around 58) chose their own "organizer," who was then expected to travel once a week to the academy for training. The extension service was restructured so that a "model farmer" chosen by the villagers became the point of contact. He too was given travel money to visit the academy weekly for training and to obtain agricultural inputs. See Millikan and Hapgood, pp. 110–111.

[49]Kahn, pp. 14–15.

important. Increasing agricultural production was seen as only one of a number of goals required to improve the life of the rural peasant. In addition to the programs on the agricultural cooperatives and the extension service (which led to the creation of local training and development centers), there were programs promoting irrigation, public works, rural education, and a special program for women.[50]

The turmoil associated with independence has had some effect on the Comilla project. The driving force behind the project, its director, A. H. Khan, had to return to Pakistan. Also, the work of the academy has become somewhat more politicized in the last couple of years. Whatever the eventual impact of these and other disruptive forces (disastrous flooding, for example), it can be noted that while generally successful in terms of the goals of rural development, a firm environmental base is still lacking in the program. Stevens notes that even in the agricultural sector there were some problems:

> Social scientists working in rural areas of developing nations are at a great disadvantage if they lack easy access to high quality agricultural and other technical knowledge. This is because productive new technology is central to most effective rural development programs whether in agriculture, nutrition or health. Due to the organizational requirements of these various technologies, considerable knowledge of the appropriate technology is essential for effective work in the social sciences in rural areas.[51]

If this is the case for conventional agricultural and health technologies, then the task of developing appropriate intermediate technologies and combining them with institutional innovation requires a new imaginative step. Also, as the reference to flooding suggests, there are often national (and in this case international) policies required to minimize the risks and damage associated with such large-scale environmental processes. For such county-sized innovations to be successful in the longer-term, they have to operate in a sound macroenvironmental milieu.

The Puebla project in Mexico offers a number of contrasts to the Comilla project—suggesting that reasonably successful projects can evolve from rather different approaches as long as they use a fairly comprehensive approach and tailor it to local conditions. The 47,500

[50]The latter included training in family planning, child care, literacy, sewing, spinning, poultry raising, sanitation, etc. See Stevens, pp. 14–16.

[51]Stevens, pp. 39–40.

farm families occupying some 323,000 acres in the Puebla region are mainly dry-land maize farmers as contrasted to the irrigated rice farmers of Comilla.[52] The project was financed and largely managed by the Rockefeller Foundation as compared to the official Pakistani backing and financing given Comilla. Finally, the orientation of those running the project was much more agronomic than the social science orientation of those managing the Comilla project. Since the Puebla project was operating largely outside normal government agencies, the director, Dr. Leobardo Jimenez, spent some time in the early stages explaining the project to government leaders in Mexico City. However, when it came time to communicate with the peasants, the team ran into difficulties. Initially, they sought basic information from selected peasants through a questionnaire. Since "this approach often earned them suspicion and mistrust," they soon switched to calling on village leaders, explaining the program to them, and asking that they call meetings to explain it to the peasants.[53] Then, two or three local "cooperators" were selected by the local peasants to work with the program.

The Puebla program has had two major thrusts: carrying out research and demonstration work, and seeking better support and delivery systems for inputs. Since the Puebla region is largely one of dry-land farming and since there was no local experiment station, the project staff divided the region into five zones to account for variations in soil, elevation, rainfall, etc. Different combinations of planting dates, maize varieties, and plant density were tried on local farms. Participating farmers received free fertilizer and seed, plus the harvest (minus enough for future breeding work).[54] The researchers found that the HYVs from CIMMYT did not yield as well as local varieties developed over the years by a dairy farmer, Sr. Salvatori. His varieties have the additional advantage of not being hybrids, so that next year's seed can be taken from this year's crop rather than having to be

[52]Most of the details on the Puebla project are drawn from Carroll Streeter, "Reaching the Developing World's Small Farmers," *Working Papers* (New York: Rockefeller Foundation, 1972), and Ralph W. Cummings, Jr., "Review of Plan Puebla," (New York: Rockefeller Foundation, 1973). (Mimeographed.)

[53]Streeter, p. 4.

[54]Ibid., p. 5.

purchased anew each year. The critical variables involved in increasing production were found to be the planting dates (the best dates were worked out according to elevation and soil to maximize chances of water exposure at the critical growing phases), an increase in the number of plants per acre (from 15,000–20,000 plants per acre up to 50,000 plants per acre), and an increase in fertilization (with zonal variation) to support the greater plant density.[55]

The second thrust of the Puebla project has involved the communication of improved agronomic practices and credit information through two locally produced films, the establishment of local groups with self-selected leaders to serve as cooperatives for gaining credit, and the use of these groups as a conduit for pooled fertilizer orders.[56] While there have been some changes in policy on the part of various government agencies in regard to credit, fertilizer, and grain purchasing, the main impact of the project has been to encourage Mexico State to adopt a similar but geographically more extensive program. The results within the Puebla region have been an increase in maize production of some 30% between 1967 and 1972 (even though only some 25% of the area's acreage was farmed by "cooperators"), an increase in average family income (with more of the increases occurring at the bottom end of the economic scale) and an increase in the average labor required for cultivating a hectare of maize.[57]

In reviewing the Puebla project, one can see that it continued the Rockefeller emphasis on increasing agricultural production—although here among dry-land peasants, rather than irrigated farmers. The assumption was that increased production would improve rural well-being. The limitations of using primarily an agronomic approach have been recognized by Cummings:

> One general conclusion . . . was the feeling that much more attention has been paid to the technological aspects of the program (work of very high quality) relative to socioeconomic research on what motivates the people in the area, what limitations they face, etc. . . . In this respect, some useful

[55]Ibid., p. 6, and *President's Review and Annual Report* (New York: Rockefeller Foundation, 1968–1972).

[56]Ibid., pp. 6–7.

[57]See Cummings, pp. 8–12, for the various statistics and the measures used.

> lessons might be learned from the Comilla Project . . . its staff was composed primarily of social scientists, it first asked what the people wanted/needed and then implemented a wide range of programs to attempt to improve these situations. . . . [58]

It should also be noted that the Puebla project had less interest in the full range of rural development programs: nutrition, women's programs, rural education, etc. Thus, while innovative in trying to move the green revolution approach from the experiment station to the field and, more importantly, from a focus on irrigated to dry-land farming, the project did not concern itself with the broader aspects of rural development. Even on the natural science side, it made no attempt to go beyond the narrow, production-oriented specialists to include generalists who would assess larger-term environmental consequences. While obviously a private foundation is not in a position to sponsor and carry out as comprehensive a project as a government can, Puebla could have included research and training on nutrition and could have included some environmental scientists (not to mention a few social scientists).

Medium-Scale Attempts: Japan and Taiwan

Turning to examples of medium-scale rural development, a number of national policy dimensions become more visible. The examples chosen—Japan and Taiwan—illustrate these, since both operate as independent states. One striking aspect, particularly in Japan, is the contrast between the conservation-oriented policies followed in agriculture and forestry and the environmentally destructive ones followed in the industrial sector. The question that arises here is whether, after gaining much of its initial sustenance from agriculture, industry will now and in the future so poison the environment that agriculture will become its casualty. Taiwan appears to be somewhat better off, but the lack of studies on environmental interactions between the industrial and agricultural sectors makes any projections difficult.

[58]Ibid., p. 17. Much stronger criticism of the project has come out of a recent study of two villages in the area. The technocratic approach of those involved—essentially their imposing a preconceived package from above without fully explaining the purposes or later on giving out the results—was said to have led to serious communication gaps between the researchers and the cooperating peasants. For details see Pearse's, *An Overview Report,* Chapter 11.

A couple of common characteristics in the development of agriculture in both countries can be noted. The problems of soil conservation, watershed management, irrigation, and drainage were handled on the local level through the use of labor-intensive techniques and small-scale projects because the river valleys are short and the coastal plains narrow, a geographic advantage not present in many parts of Asia.[59] Another common experience was that of effective land reform carried out since World War II. In each case, there was significant external pressure in bringing about the reform, but perhaps more importantly, in each case there was a great deal of local involvement in working out the details.[60] Also, in each, a fairly low ceiling was established on the number of acres even a resident landlord could own.[61]

There are a number of other aspects of Japan's agricultural history that make it atypical of current approaches to agricultural development. A brief review will show that a number of them are compatible with the contextual approach. Since Japan has some of the highest crop yields per acre in the world, the productivity of their approach is clear. The roots of this productivity—as in the United States—are deep, going back some 100 years to the policies and institutions established during the Meiji restoration. These were complemented by cultural traditions of self-discipline and self-sacrifice for the good of the larger group. Although there were some false starts—the soil erosion problems created with the removal of land-use restrictions (see Chapter 5) and the failure of attempts to introduce large-scale Western farming —the amount of successful innovation is impressive.[62] There was an analysis and mapping of soil types. Feudal restrictions on land use, labor movement, and choice of occupation were removed. Education was given a practical and technical focus, and agricultural colleges and research stations were created. Procedures and instructions were developed to adapt seed varieties and fertilizer dosages to local soil and

[59]Bruce F. Johnston, "The Japanese 'Model' of Agricultural Development: Its Relevance to Developing Nations," in *Agriculture and Economic Growth: Japan's Experience*, edited by Kazushi Ohkawa, Bruce F. Johnston, and Hiromitsu Kaneda (Princeton, N.J.: Princeton University Press, 1969), p. 96.

[60]See Dore and Koo.

[61]In Japan, the limit is 2.5 acres. In Taiwan, it varies according to soil quality. The abolishment of absentee-landlordism meant that owner-operated farms went from 54% to 92% in Japan and from 60% to 80% in Taiwan. Millikan and Hapgood, p. 98.

[62]The following is based on Johnston, "The Japanese 'Model' of Development . . . ," pp. 60–61.

climate. Local "agricultural improvement societies" were encouraged. In all of this, it "is apparent that political leaders and government officials at all levels regarded agricultural improvement as a matter of great importance and gave significant support to the work of agricultural research and extension personnel."[63]

Most of the above measures were consistent with the contextual approach suggested here. The results—increased productivity combined with capital-saving, labor-intensive approaches—meant that Japan was able to use agriculture as the base upon which to build its industry. However, as the balance between the agricultural and industrial sectors has shifted, there have been not only economic and political dislocations but the potential problem—raised above—that industry may now have encapsulated the agricultural sector. The political dislocations associated with this shift—particularly during the interwar period—manifested themselves in the growth of an agrarian ideology that has some interesting parallels to the populist movement in the United States.[64] The psychological shock of defeat in World War II, the presence of occupation forces bent on democratizing Japan, and the accumulated abuses of absentee landlordism all contributed to the thorough-going land reform which followed.

A century's work in developing small-scale, locally adapted modes of agricultural production has made Japan today one of the two basic potential models for other developing countries:

> Most developing countries face a basic issue of agricultural development strategy that can be crudely defined as a choice between the "Japanese model" and the "Mexican model." . . . In essence, the contrast between the Japanese and Mexican approaches to agricultural development lies in the fact that the increase in farm output and productivity in Japan resulted from the wide-spread adoption of improved techniques by the great majority of the nation's farmers whereas in Mexico a major part of the impressive increases in agricultural output in the postwar period have been the result of extremely large increases in production by a very small number of large-scale, highly commercial farm operators.[65]

---

[63]Ibid., pp. 61–62.

[64]See Thomas R. H. Havens, *Farm and Nation in Modern Japan* (Princeton, N.J.: Princeton University Press, 1974), pp. 313–317. Japanese agrarianism was, however, much less interesed in agricultural policy *per se;* rather, it used a rural utopianism to criticize increasing industrialization and capitalism.

[65]Johnston, pp. 86–87.

Even though he is careful to point out that today developing countries face various difficulties that Japan did not (population explosions, greater external pressures, etc.), Johnston leaves no doubt as to which model he feels is the most appropriate:

> There are persuasive considerations which suggest that the long-term goal of economic growth, as well as the welfare of the bulk of the population who will unavoidably remain in agriculture for some decades at least, will be far better served if agricultural development strategy is directed at raising the productivity of the existing small-scale, labor-intensive agriculture.[66]

This clearly parallels the recommendations developed herein but leaves unresolved the question of the fate of agriculture in Japan itself: Will it be strangled by pollution, industrial pressures, and energy shortages, or will a new group of innovative leaders arise to face the dilemmas of massive industrialization on a small island?

The historical development of modern agriculture in Taiwan closely parallels that of Japan—not least because Japan controlled Taiwan from 1895 to 1945. Also, there was in both a strong American influence in the postwar period. The basic measures that Japanese administrators took in Taiwan proceeded from their experience at home but were different in some regards because of Taiwan's colonial status. There were a number of infrastructural projects, primarily new roads and ports.[67] Japanese weights and measures were introduced, as was Japanese currency. Taiwanese banks and financial matters were integrated into the Japanese money market. More important for agriculture itself were the projects promoted and supervised by Japanese experts. There were land and forest surveys, the function of which was to assess the resources and to provide clear definitions of land ownership. Given the mountainous character of much of Taiwan, water conservation and irrigation projects were of great importance. The latter often also included flood control measures. These measures were buttressed by the establishment of agricultural research stations, which focused upon the development of better varieties of rice and sugar cane as well as on finding appropriate fertilizer regimes. Dissemination of new seeds and techniques was promoted by both an

[66]Ibid., p. 99.
[67]The following is based largely upon Koo, pp. 8–26.

extension service and the creation of an extensive network of agricultural and irrigation associations.

All of the above measures provided a solid foundation for the post-World War II land reforms and for the intensification of agriculture. A new institutional creation also had a major role: the Joint Commission on Rural Reconstruction. Originating in a 1945 Chinese government request for American assistance in agricultural development, the eventual result was a commission composed of two Americans and three Chinese, each with extensive experience in the other's country.[68] By the time the commission was finally established in 1948, the situation of the Nationalist regime in China was deteriorating, and the commission had only one somewhat chaotic year on the mainland before it fled, along with the Nationalists, to Taiwan. Once there, the commission and its relatively small staff (some 115 specialists) began work on a three-pronged land reform. The first prong, started in January 1949, involved improving tenancy terms by requiring a written contract of no less than six years. Even more important was the reduction of the rent tenants paid to 37.5% of the main crop yield (prior to reform it was often 50% and ran as high as 70%). The second prong, commenced in June 1951, was the sale of public land acquired after World War II from Japanese nationals and corporations. With the exception of lands set aside for the Taiwan Sugar Corporation, there was a fairly low limit on the amount any family could purchase. A preference was given to the existing tenant cultivators. The third and most important prong, the "land-to-the-tiller" program, was instigated in 1953. The lands of absentee landlords were compulsorily purchased and paid for in industrial shares and commodity bonds.[69] Resident landlords had a maximum acreage that they could retain (depending upon the land quality). The net result of these measures has been to reduce the number of large holdings, although some problems of fragmentation remain.[70]

After its work on land reform, the Joint Commission on Rural Reconstruction expanded its work to other areas, generally following the approach of initiating projects with local consultation; these proj-

[68]For a detailed history of the commission, see Tsung-han Shen, *The Sino-American Joint Commission on Rural Reconstruction* (Ithaca, N.Y.: Cornell University Press, 1970).
[69]Millikan and Hapgood, p. 100.
[70]See Koo, pp. 37–43, for details.

ects, once established, were turned over to local administration. The major areas of work have been irrigation and land reclamation, crop and livestock production, forestry, fishery, rural health (including family planning), marketing, agricultural extension, and farmers' organizations.[71] As an example of local participation, extension agents are hired by the farmers' associations, which pay two-thirds of the cost.[72] After United States economic aid to Taiwan ended in 1965, the American portion of commission financing came from residual U.S. funds plus some P.L. 480 assistance. The status of the commission as the United States moves toward formal recognition of the People's Republic of China is unclear. Although the problems facing Taiwanese agriculture are in some ways similar to those of the Japanese, there is less industrialization and less pollution. In this sense, Taiwan can be seen to represent a stage of agroindustrial development halfway between Japan and the People's Republic of China, to which we now turn.

A Large-Scale Attempt: China

China is different than the previous examples in both the scale of the country and its unusual history. While there have been many different histories written of China, tracing the rise of her culture and civilization, the rise and fall of various dynasties, her relations with the outside world, etc., one of the most interesting approaches has been that of Karl Wittfogel, who attempts to outline the interactions between ecological and cultural-institutional systems. His argument is that China gradually developed a "hydraulic civilization" through its attempts to manage its environment through the construction of large-scale hydraulic works.[73] These works had two primary functions: one productive (irrigation) and one protective (flood control).[74]

[71]Hsuing Wan, "Agricultural Research Organizations in Taiwan," in *National Agricultural Research Systems in Asia,* edited by Albert H. Moseman (New York: Agricultural Development Council, 1971), p. 95.

[72]Millikan and Hapgood, p. 117.

[73]Karl A. Wittfogel, *Agriculture: A Key to the Understanding of Chinese Society, Past and Present* (Canberra: Australian National University Press, 1970). This useful summary is based upon his more extensive work: *Oriental Despotism* (New Haven, Conn.: Yale University Press, 1957).

[74]Ibid., p. 5.

In addition, through the development of the Imperial Canal, the Chinese added a third: a communications network. The administrative and management requirements of these vast systems required, in Wittfogel's view, the development of a single-centered society with a ruling bureaucracy.[75]

One of the major impacts on China of outside intervention in the 19th and 20th centuries was the decline in support for these waterworks. Wittfogel estimates that between 1850 and 1905 the percentage of government funds going for maintenance of the waterworks declined from 12% to 1.5%, with even further neglect until the establishment of the People's Republic in 1949.[76] Thus, the new regime faced not only a war-damaged country and economy but a neglected and damaged waterworks system, historically its most important infrastructural element. As the broad outlines of the various policy attempts of the new regime to cope with these problems are sketched below, a couple of major cultural supports for agricultural development should be kept in mind. First, the skills of the traditional peasant farmers have been, on average, great. Especially important has been their long awareness of the importance of plant breeding as well as the awareness of the importance of using fertilizers.[77] Another basic support has been the peasants' implicit appreciation of the value of savings, as it is woven through Chinese conceptions of the interlinking of past and future generations.[78]

The basic conception of agricultural development in China since 1949 has been based upon the intertwining ideas of a "socialist" and a "technical" transformation. Or as Mao Tse-tung stated in 1955:

> We are now carrying out a revolution not only in the social system, the change from private to public ownership, but also in technology, the change from handicraft to large-scale modern machine production, and the two revolutions are interconnected.[79]

[75]Ibid., pp. 2–6. He argues against a "feudal" interpretation of Chinese history on this basis.

[76]Ibid., p. 7. The Sino-Japanese War led the Nationalists, in 1938, to break some of the dikes on the Yellow River in an attempt to slow the Japanese armies; 2.3 million acres were flooded as a result, and irrigation was lost to another 1.7 million acres. Leslie T. C. Kou, *The Technical Transformation of Agriculture in Communist China* (New York: Praeger, 1972), p. 4.

[77]Benedict Stavis, *Making Green Revolution: The Politics of Agricultural Development in China* (Ithaca, N.Y.: Cornell University Press, Rural Development Committee, Monograph No. 1, 1974), pp. 76–77.

[78]Ibid., p. 77.

[79]Kou, p. v.

It is curious that observers of agricultural development in India have tended to stress the technical side of agricultural change (the green revolution), while observers of China have been more fascinated with the political aspects. While most observers agree that the institutional changes involved in the socialist revolution have been crucial to the agricultural development of China, fewer argue for the primacy of institutional change in India (or other developing countries) as the only way that technical programs can be expected to be either successful or beneficial in the long run.

Two major debates appear to run through Chinese attempts to effectively interconnect the institutional and technical transformations. One relates to the overall priority ranking of industry as compared to agriculture, the other to the relative priority of the "socialist" as compared to the "technical" revolution. In regard to the first debate, it is clear that industry had a much higher priority than agriculture until 1960. Agriculture was dealt with only piecemeal during the "Economic Recovery Period" (1949–1952) and as sector of only secondary importance in the first and second five-year plans (1953–1957, 1957–1962).[80] Only after the results of disastrously poor crops in 1959–1961 began to be felt (1960) was a major turnabout effected: thereafter agriculture was to have first priority and industry was to serve and assist in the development of agriculture.[81]

The second debate—that regarding the relative importance of institutional versus technical change—has been a continuous source of strife within the party. Mao generally gave primacy to the socialist transformation—arguing that only through the creation of cooperatives and then collectives could the suitable conditions be created for instituting modern agriculture. Liu Shao-chi'i, however, argued that the industrial capacity for mechanizing agriculture had to be created first; then the mechanization of farming could proceed, followed by gradual collectivization.[82] This basic difference has meant not only that the various factions supporting each view have had their ups and downs as other conditions changed but that the debates over the degree and kind of mechanization (as well as over small- versus large-scale industrialization) have been related to the political struggle for

[80]Ibid., p. 5. During the first Five-Year Plan the allocation to agriculture was 8.2%, while during the second, it was raised to 10% (p. 233).

[81]Ibid., pp. v–vi.

[82]Ibid., p. 201.

policy direction and have been important in and of themselves. For example, Mao consistently favored mechanization in agriculture because it would consolidate the collective economy and wipe out the "small producer way of thinking," while others favored chemicals and seed improvement—which are more scale-neutral.[83]

Without going into all of the various shifts in emphasis in Chinese policy relating to agriculture, it can be said that the major thrust was upon the "socialist" revolution until about 1960, with a greater concern and emphasis upon the "technical" revolution developing thereafter.[84] The "socialist" revolution involved a complex series of policies and reforms that involved both land reform and a gradual shift from cooperatives to collectives to communes. In comparison to the Soviet experience with land reform, the Chinese were able to achieve their reform with less bloodshed and without alienating masses of peasants. There are several reasons.[85] The Chinese leadership was much more rural-oriented than the urban intellectuals of the USSR. The Chinese used a step-by-step approach that drew upon the need of many peasants to pool their generally undersized plots into viable farms. Also, the timing of land reform was such that poor peasants felt themselves the beneficiaries of government policy, while in the USSR, many peasants felt that collectivization deprived them of the hard-won land reform that they had earlier carried out largely by themselves. Without going into all of the various administrative forms that have been tried in agriculture, it is important to note that small private plots have been tolerated in China and that they continue to contribute around 20% of rural income.[86]

The broad outlines for the "technical" transformation of agriculture were set out in "The National Program for Agricultural Development, 1956–1967."[87] This program, which was developed by Mao, was the beginning of a uniform nationwide crash program to try to modernize agriculture. The program was publicized through the use of eight Chinese words so that farmers could easily memorize them. They are *water*, *fertilizer*, *soil* (conservation), *seeds* (selection), *closeness* (of plant-

[83]Stavis, pp. 106–107.

[84]For a good summary of the "socialist" revolution, see Kang Chao, *Agricultural Production in China: 1949–1965* (Madison: University of Wisconsin Press, 1970), pp. 11–35.

[85]The following is drawn from Chao, pp. 51–55.

[86]Stavis, p. 57.

[87]For a full translation, see Kuo, pp. 243–262.

ing), *protection* (of plants), *implements,* and *management* (of fields).[88] Unfortunately no instructions were given to farmers on how to go about ranking and implementing these goals. In terms of national infrastructural development, the planning priorities were: first, irrigation; second, fertilization; and third, mechanization and electrification.[89]

In 1957–1958 the top-priority goal was pursued through the massive mobilization of millions of peasants to construct water conservation and irrigation systems. Because there was no effective national management of the programs so strongly encouraged by the government, great difficulties and costs resulted. Without overall management, the rapid building of new diversion channels and the moving of weirs led to a serious undermining of the flood prevention capabilities of the old networks. The new networks that were constructed also had unknown hydrologic capabilities. The disastrous floods following the heavy rains of 1959–1961 can thus be seen to be as much man-induced as "natural." Two other serious problems were related to the crash irrigation program. The great emphasis upon gravity-flow systems meant that from 5% to 20% of the available arable land was used for reservoirs and canals.[90] These often lacked proper drainage, resulting in problems of salinization. Policy corrections were instituted in the early 1960s to try to improve both situations.

Another major corrective for some of the excesses of pursuing uniform, nationwide programs has been the idea of "walking on two legs," that is, the idea that traditional knowledge, skills, and approaches should be continued and drawn upon as well as modern ones. This approach has been pursued at both national and sectoral levels. At the national level, it can be seen in the "sending down" process, whereby scientists are encouraged to spend time in rural areas and to observe and learn the peasant's problems and approaches. Equally, peasants are encouraged to participate and experiment with plant breeding, developing better tools, cultivation techniques, etc.[91] Also, the Chinese Academy of Agricultural Science has published not only modern manuals but the results of their efforts to collect "the past

[88]Chao, p. 78.
[89]Ibid., p. 79.
[90]Ibid., p. 134.
[91]Kuo, p. 14.

experiences of the masses." It has also reprinted, under the auspices of its Research Institute of Agricultural Heritage, a number of historic works on agriculture.[92] In specific sectors, there are examples of "walking on two legs" found in the joint use of modern calendars and the traditional Chinese calendar (with its associated weather adages), in the use of fertilizers (where until recently, the stress was placed on organic fertilizers), and in the making of pesticides, especially "native pesticides" derived from some 500 varieties of plants and minerals.[93]

By the early 1960s, partially as a result of the failure of the Great Leap Forward and the tremendous agricultural losses of the flood years from 1959 to 1961, the decision was made to reverse priorities to give agriculture priority over industrialization and to implement fully the technical transformation of agriculture.[94] While this decision did not settle the many disputes about how best to effect the technical transformation—especially the debates on large-scale mechanization versus small-scale localized approaches—it did mean that from this point onward, China would be committing herself more and more to the green revolution approach, at least insofar as technical measures were concerned. It should be noted that a number of important administrative changes went along with this decision, changes that in many cases were speeded up during the Great Proletarian Cultural Revolution. Three major areas can be described to suggest the importance of institutional-technical interactions. First, there was the process of "ruralizing" major administrative services and infrastructures—especially those relating to commercial and banking services. This involved changing the elitism and urbanism of many government bureaucracies—an especial target of the Cultural Revolution.[95] Also involved here was the physical relocation of agricultural colleges from urban to rural areas.[96] Second, there was a major effort to strengthen agrotechnical services. In 1963, a two-month meeting of 1,200 scientists and agricultural specialists was held to discuss planning and research strategies.[97] Finally, there were a number of educa-

[92]Ibid., pp. 15–16.
[93]Ibid., pp. 86, 97, 179.
[94]This was formalized in September 1962 at the 10th Plenary Session of the 8th Central Committee. Stavis, p. 95.
[95]Ibid., p. 157.
[96]See *Plant Studies in the People's Republic of China* (Washington, D.C.: National Academy of Sciences, 1975), p. 141.
[97]Stavis, pp. 158–160.

tional reforms involving mass literacy programs, the increase of the "sending down" approach, and (in 1964) the development of a new system of agricultural schools that, rather than serving as means for social mobility, was based on the principle of "from the commune and back to the commune."[98] Whether these measures in the long run will overcome the loss of incentive that critics like Wittfogel claim results from collectivization remains to be seen.[99] In any case, the bulk of economic improvement in China since the 1950s has been in the rural areas; rural income roughly doubled between the early 1950s and the late 1960s (from 70 to 150 yen).[100]

Current assessments of the productivity of Chinese agriculture tend to agree on the overall success of the Chinese not only in providing sufficient food for their large population but in having promoted general rural development. However, a number of potential problem areas have been identified. Perhaps most serious in the long run is the risk of genetic erosion. While the approach of seed self-sufficiency among communes reduces the risks of plant diseases spreading, the great emphasis on new hybrids that are earlier-maturing (to encourage multiple cropping) has meant a neglect and increasing loss of the old land-race populations.[101] Another "erosion" relates to the fact that most of the highly trained agricultural scientists and scholars are now elderly. It is not clear whether the new approach to agricultural education has provided, or will be able to provide, effective replacements—either in the form of a new generation of experts or through new institutional arrangements that might supply interdisciplinary substitutes for specialized approaches. One potential problem that has received little discussion in or outside of China is that of increasing regional inequalities. Given the emphasis upon the "self-reliance" of communes, provinces, etc., regional disparities could well result from a variety of causes.[102] Finally, questions relating to mechanization continue to divide the leadership in China. A key part of the new Five-Year Plan is the proposal to double agricultural manpower—

[98]Ibid., pp. 207–210.

[99]See Wittfogel, pp. 8–13. He claims that a confused comparison between agricultural and industrial production—which leads to a promotion of the large-scale "factory farm"—causes additional difficulties.

[100]Stavis, p. 54.

[101]*Plant Studies* . . . , pp. 147–159. This loss is exacerbated by the lack of a national germ-plasm preservation system.

[102]Stavis, p. 252.

primarily through mechanization.[103] One of the main points of controversy appears to relate to whether mechanization should be based upon many small and medium factories or upon big, all-inclusive industrial plants.[104] Decisions or nondecisions in these various areas will have a profound effect upon the future of Chinese agricultural production.

What, then, can be concluded about the Chinese experience in agriculture since 1949? It would appear to be an interesting mixture of adaptive and maladaptive policies—which is not surprising, given the magnitude of the infrastructural and social changes that the regime has sought. The approach appears to have been one involving trial and error, ideological predilections, and leadership competition. The most adaptive aspect involves the concept of "walking on two legs," which, while it has been filled with varying content in different fields, has involved an explicit recognition of the importance and value of traditional ways and practices. This recognition is quite in contrast to the views of the modernizing elites in most developing countries. On a larger scale, and parallel to the idea of walking on two legs, is the fundamental ideological view that rural development and change can occur only through an interaction of the "socialist" and the "technological" revolutions. While there have been leadership differences over priorities and pace, there has been agreement that both are needed. Again, this approach contrasts with most other developing countries and with the approach of most aid-donating countries—where technological revolutions (like the high-yielding varieties) are promoted as a substitute for, rather than as a complement to, major social change and reform.

The maladaptive aspects of the Chinese experience appear to be similar to many of the underlying problems associated with industrial agriculture. That is, many programs were pursued on a universal (or "mass line") basis throughout the country without adequate consideration of local conditions or problems. The sometimes rapid changes in the party line—reflecting leadership changes—often exacerbated the problems associated with this approach. These changes may relate to ideological predilections, but a good part was also the result of

[103]Victor Zorza, "Will Mao's Cupboard Be Bare?" *Guardian Weekly* 23(November 1975):11.
[104]Ibid.

the leadership's attempt to "catch up" with other major powers. This attempt was most prominent during the early period, when industrialization was given top priority. Another problem in the early period was the tendency to copy blindly what were then seen as more advanced Soviet approaches and technologies. For example, in the early 1950s, efforts were made to promote the theories of Soviet geneticist Trofim D. Lysenko.[105] During the same period, there was a plan to introduce some 6 million large Soviet-style plows. These were much too large for most Chinese plots, and eventually less than 100,000 were distributed.[106] These sorts of problems ended with the Sino-Soviet split, which had an additional healthy result: Chinese researchers had to be given greater responsibilities and respect.

While many have suggested that China be viewed as a model for rural development, the above, plus more general cautions, should warn against any direct copying (something the Chinese themselves say). Those who would learn from China should look closely at the various parameters that the Chinese have struggled with and the various approaches that have been attempted—some with more and some with less success. The fundamental interactions between social and technical transformations deserve a great deal more study. Also, with the exception of the few people in the West concerned with "intermediate" or "appropriate" technologies, the Chinese have had long-standing and explicit discussions on the variable social impact of different kinds of technologies. In a sense, these discussions can be seen as "smaller-scale" versions of their overall emphasis on the mutual interactions between the social and technological transformations. And, of course, there is an important time dimension: the Chinese "model" has changed rather significantly in the past and can be expected to do so in the future. Or, as Stavis puts it, "China's experiences do not constitute a simple model which can be adopted by other countries. They may not even be suitable for the China of the future."[107]

[105]Kou, pp. 154–156. After Stalin's death and Lysenko's removal from office, the Chinese decided that genetic questions should be seen as scientific/technical questions, where there would be no party line.

[106]Ibid., pp. 192–193.

[107]Stavis, p. 274.

LESSONS FOR PLANNERS AND POLICY MAKERS

A variety of individual recommendations has been discussed in the chapter. The "strategic" recommendations can be summarized as follows:

1. There is a need for new approaches—"alternative data systems"—to collect, organize, and synthesize data on natural resources in such a way as to be locally useful.
2. The political and organizational aspects of data collection must be kept constantly in mind.
3. Questions regarding the appropriate scales of agricultural production need to be reassessed in terms of the local limiting factors, in terms of energy efficiencies, and in terms of social and environmental impact.
4. There is a need to examine other scale and locational aspects of agriculture, especially those relating to infrastructure, local institutions, and land use.
5. The quantitative emphasis on production must be replaced by a larger qualitative emphasis on the well-being and the development of the rural masses.

The "policy" recommendations included the following:

1. Policies aimed at promoting locally adapted and sustainable agriculture must be based upon a much greater conservation of basic resources—water, energy, soil, and plant varieties—than has been the case in conventional approaches.
2. Policy makers and advisers need to develop a much greater awareness of the interactions between physical and social systems.
3. In contrast to conventional development approaches that seek to integrate the peasant fully into a proposed modern agribusiness complex, the goal should be to make the marginal peasant more genuinely self-sufficient (in nonmonetary terms) while at the same time giving him a small cash surplus that can be used for education, etc.

Three more general themes that emerge from the above recommendations and brief case studies are the importance of scale, the importance of spatial location, and the importance of the time dimension. The first of these, scale, was illustrated in a number of ways. Perhaps the most important point to emerge was that differing perceptions and standards of evaluation are used at different scales. When an analyst is examining a local project like Comilla or Puebla, he tends to take for granted a number of points present or absent in the national framework. Equally, when examining individual states, there is a noticeable difference between discussions of a small state like Taiwan and a large state like China. Even in the case of the latter, a number of important global elements are either assumed or ignored. In each, somewhat different units of analysis tend to be used.

It is quite curious that in spite of the differences in perceptions and standards of evaluation that can be noted as scales shift, the conventional understanding is that generally scale is neutral. There are exceptions, of course. Economists have cultivated the concept of "economies of scale," although at the same time they tend to stress the neutrality of the market mechanism (in terms of size), the neutrality of technologies, and the neutrality of institutional structures (as a function of size). The lesson that planners and policy makers need to learn here is not only to question the assumption that "bigger is better" but to question the conventional wisdom that technologies, market mechanisms, and institutions in general are size- or scale-neutral. This was particularly stressed in regard to the question of the appropriate size of the production unit—which was shown to vary according to the local limiting factor (whether environmental, cultural, or institutional).

The importance of spatial locations and relationships is another basic factor that has tended to be ignored or only selectively recognized. It was discussed in terms of the physical location of schools as well as of irrigation systems but obviously needs to include all relevant factors for agriculture and rural development. The idea of "intensive development zones" is one way of trying to integrate environmental, cultural, and institutional factors. What is required of planners and policy makers here is first an analysis of the basic resources, cultural patterns, and institutional structures (all in terms of the larger requirements for sustainable agriculture and rural development), and then an attempt to develop a local pattern of infrastructure, institutions, and policies that promotes an integrated program of sustainable development. As has been suggested several times, it would appear clear that the primary development goal has to be rural and agricultural rather than industrial if such an approach is to work. Only a few of the poor countries—like China—have genuinely given rural and agricultural development top priority.

The great importance of the time dimension has been stressed throughout the book. Not only does the choice of different time-frames require different units of analysis, but they illustrate different types of problems. For example, assessments of the limiting factors in a given region may well vary according to the time-frame chosen. This point is particularly important to keep in mind in conjunction with discussions of scale variations. Also a consideration of larger time periods helps to illustrate the importance of institutional changes or reforms, such as land reform. Equally it is extremely valuable—so long as one recognizes the nonneutrality of technologies—in helping to trace the interactions between institutional and technical changes. The Chinese case is particularly illustrative because they have explicitly recognized the importance of these interactions. The major lesson for planners and decision makers here is that attempts to promote technical transformations alone (because institutional reform is seen as too difficult) may well lead to worse results than either no reform or institutional reform alone. Clearly, the analysis here suggests that complementary technical and institutional reforms would be best, but unfortunately there are few examples or case studies of such complementary interactions to draw upon.

The path that needs to be followed if we are to go beyond current conceptions of rural development has been largely traced out. One final aspect that needs to be stressed in regard to the case studies is the importance of global influences and parameters. All states are ultimately dependent upon moderate climate conditions; the availability and purity of such basic resources as water, soil, and air; and the general viability of global ecosystems. In addition, it must be noted that two of the three states examined here—Japan and Taiwan—are dependent in significant ways upon external resources and trade, especially imported petroleum, fertilizers, and grain. The amounts of protein represented by fish caught and imported are also significant. Thus, these states are far from being self-sufficient (not that this is necessarily an ultimate goal) and, more importantly, are dependent upon resources that are either limited or threatened.

These sorts of uncertainties suggest the need for strengthened international agencies to deal with the various global and strategic dimensions of the environment and its single most important sector: agriculture. Also they suggest that national planners and decision makers need to rethink their domestic and international priorities if either the global system or the national subsystems are to become adaptive enough to survive the various stresses and pressures that have been delineated throughout the book.

7

Hindsight, Insight, and Foresight

After a rather extensive exploratory survey of global agriculture (past, present, and future), it is time to integrate the main themes. A major theoretical theme has been that it is only through new modes of analysis that take full account of context that the dynamic reality of interacting physical, biological, and social systems can be approached. The basic elements involved in such contextual analysis include a conscious sensitivity to the importance of variation over space and time and the resulting distribution patterns of a wide range of things (soil types, vegetation, rainfall, infrastructures, institutions, land ownership, wealth, etc.). The neutrality of a number of concepts and practices has been questioned both analytically and through examples. Thereby, the Western origins, character, and impact of a number of concepts and technologies have been shown. Analytically, the simultaneous use of three time-frames (evolutionary, developmental, and policy) helped to illustrate various ways in which our understandings of agriculture have been distorted by stereotypes, selective perception, and universal assumptions. How pervasive these are was shown by reviewing various specialized disciplines (themselves a product of modern Western views) and their distortions, lacunae, and value predilections. Only by understanding the general cultural context of modern Western science as well as the historical biases of each major discipline can there be any hope of gradually building a synthesis of the biological and social sciences.

The main thrust of the book has been to attempt such a synthesis as it relates to global agriculture. Use of the three time-frames has involved a process rather like using a zoom lens on a movie camera, moving back and forth from general panoramas to intermediate shots to close-ups. As one moves from one level of analysis to another (as from one focal length to another), there is an inevitable trade-off between scope and detail. More importantly, as different phenomena and processes become visible, different kinds (or levels) of variability can be noted. Their interlinkages and relative importance can then be assessed and the true dynamics of change more closely approached. Using the photographic metaphor, one can see how the image of an individual building (with all bricks visible) fits in with the image of its village (where the relationship of the building to others is visible) and how that fits into a panorama of the general countryside showing both the village and its environs. In this way, one can avoid one of the major sources of error in much contemporary thinking: the assumption that "for all practical purposes," many phenomena and processes are constant, when in point of fact they are quite variable.

When contextual analysis is applied to global agriculture, a very unconventional three-level mosaic results that suggests, among other things, the need for a major reevaluation of our understanding of agricultural history. In particular, there is need to review critically the general tendency to project Western and temperate zone conceptions upon the rest of the world. Also, much more work needs to be done to analyze, describe, and incorporate evolutionary and developmental parameters into our thinking. Even the modest degree to which this has been done in this book has resulted in a very different set of standards for evaluating agriculture than is typical of either economists or developmental experts. One also arrives at a much better understanding of the momentum of the past, whether embodied in the form of infrastructures, institutions, or generations. The latter, it was suggested, have been almost totally neglected by social scientists. The examination of the "development plays" of the modern era suggested that the green revolution has had many precursors that paralleled on a smaller scale its weaknesses and risks, particularly the attempt to transfer practices and techniques that are thought to be neutral but in point of fact carry large amounts of cultural and environmental baggage with them.

However, much more than hindsight is required. As the metaphor of "development plays" explicitly suggested, the present must be understood in terms of its own basic context—which is different than that of previous centuries. As we have incrementally built societies that are structured around and dependent upon large-scale institutions, infrastructures, and technologies, there has been an increasing need to examine the future. This is because these large-scale creations, while powerful, also are rather inflexible and slow to change; they have a great deal of momentum, which, if wrongly directed, risks carrying industrial societies down potentially catastrophic paths. This basic fact plus the risks of accumulated and ever-increasing environmental degradation has dictated a widespread interest in the future.

Contextual analysis has been particularly helpful in illuminating some basic weaknesses in most current approaches to the future. In terms of the critical question of priorities, the time-frames help sort out different levels of priorities. I have argued that protection of those fundamental life-support systems associated with the evolutionary time-frame is of top priority if we are to have any hope of keeping our evolutionary options open. Use of the evolutionary time-frame not only helps in sorting out priorities but also helps identify which processes and systems are global and basic. Using the developmental time-frame, one sees that the industrial and nonindustrial countries will have to employ rather different strategies if they are not to exceed their respective limits. Most conventional proposals for the future suffer either from a tendency to deal with only one sector in isolation from what is happening generally throughout a society or from an inability to deal with longer-term parametric changes. The latter criticism applies to regional and global computer models, which, even though they represent a significant advance in terms of integrating the interactions of various sectors, still necessarily have a single basic structure. Another weakness of most models of the future—which flows directly from our fragmented and partial views of the past—is that they seriously neglect or underestimate the importance of environmental factors, especially biological ones.

All of these weaknesses are especially visible when one explores the future of agriculture. Most proposals for dealing with agriculture (even if they distinguish between the needs of the rich and the poor countries) tend to involve variations on current themes in industrial agriculture. The use of contextual analysis suggests that there must be a thorough reconceptualization of approaches to agriculture and that without it we run a genuine risk of catastrophe in the longer-term. Hopefully, some of the basic building blocks for reconceptualization have been developed in this work. However, it must be recognized that better analysis alone is far from sufficient to deal with the unprecedented situation we find ourselves in. As indicated above, the momentum of our infrastructures, institutions, and technologies is greater than at any other point in history. This means that extremely long lead-times are required to bring about any meaningful changes—even after the problems have been recognized and some degree of political consensus has been developed regarding "solutions." In the past, the United States has tended to use a crisis approach to any pervasive problem. Unfortunately, the lead times required today will inevitably doom this approach to being too little too late. The dilemmas involved have been painfully apparent in our attempts to deal with the energy crisis. The role of various vested interests—economic, academic, and intellectual—make the political task of transforming our society and its supporting structures very difficult. If we are encountering such difficulties in the energy field—where there are huge amounts of waste that should be easily conservable—what will happen when we have to make basic changes and conserve in "leaner" areas? Some have concluded that the above difficulties and dilemmas will lead to the demise of the great experiment in democratic self-government of the past several centuries; that only some form of authoritarianism or dictatorship will be able to impose the measures required for survival.[1] Even though a strong case can be made that dictatorial regimes will be even less able to deal effectively with such problems in the longer term, this does not greatly

[1]See especially Robert L. Heilbroner, *An Inquiry into the Human Prospect* (New York: Norton, 1975).

reduce the risk that such regimes will be installed—either by fearful elite groups or by popular demands that something be done.[2]

Clearly there are a number of prerequisites if the gloomy situation described above is to be avoided. First, there must be better and more accurate analyses of our past and present situations and of the extent of our future possibilities. Next, there is a need for a new vision of society and its goals, especially as they relate to the future. This must ultimately become broadly based and broadly shared. Karl Mannheim distinguished between ideologies (rationalizations of and defenses for the status quo) and utopias (visions of the future that require basic or radical changes in the status quo). Most of the current ideologies of industrial society (which naturally include a view of the future) are outworn and rigid—and it is the sensing of this that makes the more perceptive elites in industrial society gloomy. The less perceptive—the technological optimists—tend to promote more and more rationalistic and technocratic ideologies, not realizing that "with the relinquishment of utopias, man would lose his will to shape history and therewith his ability to understand it."[3] Mannheim's distinction also suggests that new and genuinely adaptive visions of the future will not emerge from industrial institutions and elites. This clearly appears to be the case with the three major views that are emerging to challenge industrial ideologies, whether of the capitalist or the socialist variety.

These three views are those of the environmentalists, the proponents of appropriate (or intermediate) technologies, and the radical humanists.[4] There is naturally some overlap in their views since they

[2]The classic argument on the economic inflexibility of authoritarian regimes and their resulting lack of long-term adaptability and productiveness is that of Tocqueville. See Alexis de Tocqueville, *Democracy in America* (New York: Vintage, 1954), vol. 1, pp. 237–238, 260–263; 276–279. For a modern approach, see Thomas L. Thorson, *Biopolitics* (New York: Holt, Rinehart & Winston, 1970).

[3]Karl Mannheim, *Ideology and Utopia* (New York: Harcourt, Brace, 1960), p. 263.

[4]In addition to the various environmental studies mentioned throughout the book, one should consult the thoughtful overview by William Ophuls, *Ecology and the Politics of Scarcity* (San Francisco: Freeman, 1977).

The major approaches to appropriate technology can be found in the following works: Ernst F. Schumacher, *Small Is Beautiful* (New York: Harper, 1973); David Dickson, *Alternative Technology and the Politics of Technical Change* (Glasgow: Fontana/Collins, 1974); Nicholas Jéquier, *Appropriate Technology: Problems and Prospects* (Paris: OECD Development Centre, 1976); David Elliot and Ruth Elliot, *The Control of Technology* (New York: Springer-Verlag, 1976).

The roots of radical humanism can be traced back at least a century; the most

have many of the same goals and opponents; however, their views have grown up separately and there is clearly a need for attempts at synthesizing their work. One common theme around which such a synthesis might be built is the great concern of each to understand the real time, place, and scale parameters of the various technical, natural, and human processes they are interested in. That is to say, each has a great concern for context. Also, each seeks to understand the needs and limits of real people and environments and to adapt ideas and technologies to them (much in contrast to most conventional approaches).

A preliminary comparison of the three would suggest that each has something to learn from and offer to the other. The appropriate technology (AT) groups appear to have the most to learn about time past: about geologic, evolutionary, and successional time from the environmentalists and about the history of mankind from the humanists. However, when it comes to an examination of and concern for the future, particularly how infrastructural patterns and environmental trends limit the range of future possibilities, the humanists have some things to learn from the other two groups. As regards the search for place, the AT groups could benefit from an appreciation of the importance that environmentalists give to specific ecosystems and species as well as the stress that humanists give to cultures and their diversity. The AT concern with the specific ways in which technologies both derive from and shape different cultures and environments could add a valuable dimension to the rather general understandings of the other two groups in these areas. Also, in the search for proper scale, the AT groups, with their concern for technology on a human scale, can add some useful specifics to the more general concerns of the environmentalists and humanists.

Whatever their differences and shortcomings, the important point to note is that new visions of the future are emerging in these three areas as the result of vigorous grass-root activity and debate. Their new visions represent a fresh but somewhat harried attempt to find alterna-

---

prominent contemporary writers in the field include Lewis Mumford, *The Myth of the Machine: Technics and Human Development* (New York: Harcourt, Brace, and World, 1967) (to mention only one of his many works); Ivan Illich, *Tools for Conviviality* (New York: Harper & Row, 1973); Theodore Roszak, *Where the Wasteland Ends* (Garden City, N.Y.: Doubleday, 1972); and Wendell Berry, *The Unsettling of America: Culture and Agriculture* (San Francisco: Sierra Club Books, 1977).

tives to what are seen to be the basic flaws of industrial society. These groups are now actively seeking answers to the fundamental questions they have raised regarding modern society. Are there limits to growth? If so, how can society be organized to live within them? How can we preserve the quality of environments and societies threatened by various kinds of degradation or destruction? Can progress be defined only in material terms? What are the real social and environmental costs of large-scale operations? Can decentralization help reduce alienation and dehumanization? These and many other questions permeate the ongoing discussions, whose liveliness and depth certainly are a hopeful sign; however, it must be remembered that important as a new vision of the future is, it is still only one of several prerequisites for transcending our current dilemmas.

Another prerequisite relates to the need for social innovations as well as a range of transforming strategies that will enable us to get from here to there. The question of how to achieve actual implementation of their visions has always been a problem for utopian thinkers or for radical reformers; today the situation we find ourselves in has immensely complicated the task. Clearly, several decades (at a minimum) will be required to bring about any of the goals sought by the three groups. Groups promoting such goals as energy conservation or decentralization must face several structural dilemmas that flow directly from the complicated and integrated infrastructures and institutions of industrial society. Politically, it is precisely the large economic interests (which will have the most to lose from system transformations) that are in the strongest position to block or delay changes. In this regard, the difficulty of achieving transformation in the industrial world parallels very closely that of land reform in the developing world. Transforming strategies must therefore have a strong structural component, must be long term in their conception, and must try to avoid numerous political pitfalls. Such strategies will have to be built upon the twin pillars of a new holistic science (which can effectively analyze long-term interrelationships) and a new and alternative vision of the future. Only with such foundations can transforming strategies have a chance of challenging the legitimacy of the *ancien régime* while building the political support necessary to restructure institutions and infrastructures. It should be noted that this process is rather unlikely to occur through any classical political revolution. The Chinese case, sometimes seen as an exception, offers little guidance to the industrial world; also, even in

China, the battle between those seeking greater industrial centralism and those seeking rural self-reliance is far from over. It would appear that the types of change needed will more likely emerge from a broad range of "minirevolutions" occurring at the grass-roots level. However, some way of transforming our basic infrastructures must also be found; otherwise the various experiments and minirevolutions risk ending up as small islands of sanity in an ever-rising (and increasingly polluted) industrial sea.

It is here that the final prerequisite becomes obvious. There will be a great need for *adaptive intellectual and political leadership.* Clearly, this will be required throughout all sectors of society, not just at the national and international levels. *Adaptive leadership*—something akin to what has traditionally been called *statesmanship*—involves a clear awareness of the longer-term needs of a group, some idea of the kinds of approaches or strategies that offer a good chance of bringing about whatever changes are necessary to adapt the group to the changed requirements of the future, and the patience to pursue those strategies through education, persuasion, and *genuine politics.*[5] Clearly, the intellectual and psychological demands of such leadership are great. Intellectually, an uncommon combination of wisdom and practical experience is demanded. It may be that this mixture will emerge only from the crucible of events. However, our own history suggests the value of vigorous intellectual training prior to plunging into the maelstrom of events (witness both the Founding Fathers and Abraham Lincoln). One must ask where tomorrow's leaders are to receive the kind of intellectual education appropriate to contemporary circumstances and dilemmas? Unfortunately, it appears that they must individually try to put together an education that is holistic and capable

[5]Ophuls, p. 233, reminds us that ". . . ultimately *politics is about the definition of reality itself.* As John Maynard Keynes pointed out, we are all the prisoners of dead theorists; the ideas of John Locke, Adam Smith, Karl Marx, and all the other philosophers of the Great Frontier in effect define reality for us. Before we can even see what the problem is, we must tear off their fetters on our imagination. To put it another way, normal politics is indeed 'the art of the possible'; it consists in working as best one can for valued objectives 'within the system'—that is, inside the current political paradigm. However, politicking (to give it its true name) is only one part of politics, and the lesser part at that; in its truest sense, *politics is the art of creating new possibilities* for human progress. Since the current system is ecologically defective, we must direct our concrete political activities primarily toward producing a change of consciousness that can lead to a new political paradigm" [italics in the original].

of dealing with long-term problems. The specialization of our current educational systems mitigates against such an effort, but perhaps for those who succeed in spite of the obstacles, the lessons in how to overcome specialized thought and institutions will be valuable in their pursuit of similar but larger policy goals.

If one believes, as suggested above, that democratic systems offer greater long-term potential for adaptive social responses than do authoritarian systems, then the task of adaptive leaders takes on a twofold dimension. First, such leaders need to recognize that they must ultimately persuade the members of their society to voluntarily reduce their wants and desires—especially in regard to material matters. Naturally, there can be carrots and sticks, but more important, such leaders have to overcome a number of intellectual obstacles and economic vested interests to open up our current industrial oligarchies and make them more genuinely democratic. The prerequisites for an effectively operating democracy have been seriously weakened in the past century. Democracy assumes that the citizens have sufficient data available to make informed judgments and that their judgments have a real impact on the governmental system. Rarely does a citizen have sufficient information to make an informed judgment on policy issues today. One difficulty relates to the ability of the various large institutions in our society to edit, manage, or in other ways control access to and the dissemination of critical pieces of information. Perhaps even more serious is the fact that modern society is so structured that cause and effect are much more greatly separated in time and space than ever before. In the 19th century, if a tannery was polluting a river, both the cause and the effects were observable locally and could be dealt with there. Today, an industrial plant several hundred miles away can pollute a river with very-difficult-to-detect compounds whose full toxic effects may not become known for decades. In such circumstances, few citizens are able to make informed judgments or to influence policy much. While clearly some interim form of protection must be sought for such situations—usually in the form of more governmental involvement or bureaucratic supervision—the longer-term goal should be toward a greater decentralization. Such decentralization must involve not just jurisdictional lines but institutions and infrastructures. Otherwise there will be no movement toward localizing cause-and-effect relationships and making them more controllable.

The above should not be read to mean total decentralization—some sort of return to medieval fiefdoms. Rather, it is an argument for a new type of federalism that goes beyond questions of governmental units, their respective powers, and their relation to individuals and private organizations. Any new federal theory will have to deal with questions of how environmental, societal, and technological systems interact and how they can be either intermeshed or balanced off against each other so that society is able to achieve important social and environmental goals while at the same time remaining flexible and adaptive. It may well be that the basic framework required for greater decentralization will have to be achieved or imposed through some sort of centralized system, rather like classical antitrust theory. The latter, however, illustrates the weaknesses and failures of a purely legal approach to problems of overcentralization. Thus, adaptive leadership will involve several ambiguities and paradoxes: a general predilection for decentralization combined with the necessity to use centralized tools on occasion; the psychological stress and frustration of living in the current system with the knowledge that it may take decades to transform it, while at the same time seeking to develop and refine knowledge of genuine alternatives. The challenges and risks are imposing: on the one hand, premature action risks backlashes, which would reinforce authoritarian tendencies; on the other, long delays may result in leaders' losing patience and dropping out to cultivate their own private gardens.

Lest the demands of adaptive leadership appear overwhelming, it should be repeated that such leadership must be widespread: the logic of the situation as well as the values of decentralization require it. Thus, one should expect mutual encouragement and reinforcement among adaptive leaders. Beyond that, it is well to keep in mind Joseph W. Meeker's basic point that the values of the comic tradition are much more ecologically based and adaptive than those of the tragic tradition, where the hero's quest for self-knowledge and greatness often lead to disastrous consequences.[6] Unfortunately, many ecologists and environmentalists have a long way to go before they develop a true comic sense (of course, one can ask whether technocrats even possess the potential for developing a comic sense). In any case, it is a very human

[6]Joseph W. Meeker, "The Comedy of Survival," *The Ecologist* 3 (June 1973):210–215.

enterprise we are all engaged in, even if many of its instrumentalities are nonhuman.[7] Finally, it must be stressed that the human context today is basically different than that of previous ages because while we have always been part of nature and evolution (even if we have not always perceived this), we now are in a position to strongly influence both natural and human evolution.

"Influence" is not to be read as "control"—even if most technocrats and technological optimists read it that way. Here again, one finds basically different views of the world, one emerging and one dominant. Rather like such proponents of political nonviolence as Ghandi and Martin Luther King, Jr., the proponents of a less violent approach to nature must consciously avoid adopting the basic methods and thought patterns of those they are seeking to replace. Perhaps one place to look for some guidance and inspiration is to the person around whom so much of this book has revolved: the peasant. In many ways he is similar to us: he has gaps in his knowledge, he tends to be culturebound, and many of his views of the world are filtered through various stereotypes. His concerns tend to be short term, and he may be driven to unecological practices by external pressures. However, in some very basic matters, we would do well to learn from him: he recognizes that he must work cautiously within the larger natural environment surrounding him; he knows that his margin of survival may be slim and that conservation, careful use and recycling of materials, and the mimicking of natural systems can greatly aid him. Yet in spite of his rather stark lot, *he perseveres.* Will modern man be able to do as well as he enters a period of fundamental changes and difficulty?

[7]For a brilliant elucidation of this point, see Herman Hesse's *Magister Ludi* [*The Bead Game*] (New York: Ungar, 1949).

References

Allison, Graham T. *Essence of Decision.* Boston: Little, Brown, 1971.

Aurora, Gurdip S., and Morehouse, Ward, "The Dilemma of Technological Choice: The Case of the Small Tractor." *Minerva* 10 (October 1974):433–458.

Bard, Robert L. *Food Aid and International Agricultural Trade.* Lexington, Mass.: Heath, 1972.

Barnett, Correlli. "The Education of Military Elites." In *Governing Elites,* edited by Rupert Wilkinson, pp. 193–214. New York: Oxford University Press, 1969.

Basler, Roy P., ed. *Abraham Lincoln: His Speeches and Writings.* New York: World Publishing, 1946.

Bates, Marston. "The Human Ecosystem." In the National Academy of Sciences study on *Resources and Man,* pp. 21–30. San Francisco: Freeman, 1969.

Beckford, George L. *Persistent Poverty: Underdevelopment in Plantation Economies of the Third World.* New York: Oxford University Press, 1972.

Bennett, Hugh H. *Soil Conservation.* New York: McGraw-Hill, 1939.

Berg, Alan J. *The Nutrition Factor: Its Role in National Development.* Washington, D.C.: Brookings, 1973.

Berry, Wendell. *The Unsettling of America: Culture and Agriculture.* San Francisco: Sierra Club Books, 1977.

Board of Science and Technology for International Development. *More Water for Arid Lands: Promising Technologies and Research Opportunities.* Washington, D.C.: National Academy of Sciences, 1974.

Board of Science and Technology for International Development. *Energy for Rural Development,* Washington, D.C.: National Academy of Sciences, 1976.

Borgstrom, Georg. *The Hungry Planet.* New York: Collier, 1967.

Borgstrom, Georg. *Too Many: A Study of the Earth's Biological Limitations.* New York: Macmillan, 1969.

Boserup, Ester. *Conditions of Agricultural Growth: The Economics of Agrarian Change under Population Pressures.* Chicago: Aldine, 1965.

Boulding, Kenneth E. "The Interplay of Technology and Values: The Emerging Superculture." In *Values and the Future,* edited by Kurt Baier and Nicholas Rescher, pp. 336–350. New York: Free Press, 1969.

Bozeman, Adda B. *The Future of Law in a Multicultural World.* Princeton, N.J.: Princeton University Press, 1971.

Brown, Lester. *Seeds of Change.* New York: Praeger, 1970.

Brown, Lester, *Redefining National Security.* Washington, D.C.: Worldwatch Paper No. 14, 1977.

Brownowski, Jacob. *Science and Human Values.* New York: Harper & Row, 1972.

Bryson, Reid A. "A Perspective on Climate Change." *Science* 184(17 May 1974):753–760.

Chao, Kang. *Agricultural Production in China: 1949–1965.* Madison: University of Wisconsin Press, 1970.

CIMMYT—Purdue International Symposium. *High-Quality Protein Maize.* Stroudsburg, Pa.: Dowden, Hutchinson, and Ross, 1975.

Clark, Colin. *Population Growth and Land Use.* New York: St. Martin's, 1967.

Clark, Wilson. *Energy for Survival.* Garden City, N.Y.: Anchor Books, 1975.

Cleaver, Harry M., Jr. "The Contradictions of the Green Revolution." *Monthly Review* 24(June 1972):80–111.

Clotfelter, James. *The Military in American Politics.* New York: Harper & Row, 1973.

Cloud, Preston E., ed. *Resources and Man.* San Francisco: Freeman, 1969.

"The Cocoyoc Declaration." Adopted by the participants in the UNEP/UNCTAD symposium on "Patterns of Resource Use, Environment and Development Strategies." Cocoyoc, Mexico, 8–12 October 1974. Reprinted in full in *International Organization* 29 (Summer 1975):893–901.

Cohen, Stephen P. *The Indian Army.* Berkeley: University of California Press, 1971.

Coleman, James, ed. *Education and Political Development.* Princeton, N.J.: Princeton University Press, 1965.

Committee on Geological Sciences, National Academy of Sciences. *The Earth and Human Affairs.* San Francisco: Canfield Press, 1972.

Commoner, Barry. *The Closing Circle.* New York: Knopf, 1971.

Coombs, Philip H.; Prosser, Roy C.; and Ahmed, Manzoor. *New Paths to Learning for Rural Children and Youth.* New York: International Council for Educational Development, 1973.

Corning , Peter A. "The Biological Bases of Behavior and Some Implications for Political Science." *World Politics* 23(April 1971):321–370.

Crop Productivity-Research Imperatives. Yellow Springs, Ohio: Charles F. Kettering Foundation (Book II), 1976.

Cummings, Ralph W., Jr. "Review of Plan Puebla." New York: Rockefeller Foundation, 1973. (Mimeographed.)

Dahlberg, Kenneth A. "The Technological Ethic and the Spirit of International Relations." *International Studies Quarterly* 17(March 1973):55–88.

Dahlberg, Kenneth A. "Towards a Policy of Zero Energy Growth." *The Ecologist* 3(September 1973):338–341.

Dahlberg, Kenneth A. "Ecological Effects of Current Development Processes in Less Developed Countries." In *Human Ecology and World Development,* edited by Anthony Vann and Paul Rogers, pp. 71–91. New York and London: Plenum Press, 1974.

Dahlberg, Kenneth A. "A New Approach to Evaluating Longer-Term Energy Risks." *Geothermal Energy* 4(August 1976):39–41.

Dalrymple, Dana G. *Measuring the Green Revolution: The Impact of Research on Wheat and Rice Production.* Washington, D.C.: U.S. Department of Agriculture, Economic Research Service, 1975.

Dasmann, Raymond F.; Milton, John P.; and Freeman, Peter H. *Ecological Principles for Economic Development.* New York: Wiley, 1973.

Day, Peter R. "Genetic Vulnerability of Major Crops." *Plant Genetic Resources Newsletter,* No. 27 (February 1972):2.

Dickson, David. *Alternative Technology and the Politics of Technical Change.* Glasgow: Fontana/Collins, 1974.

Doane, Robert R. *World Balance Sheet.* New York: Harper, 1957.

Dore, Ronald P. *Land Reform in Japan.* London: Oxford University Press, 1959.

Dorst, Jean. *Before Nature Dies.* Baltimore: Penguin Books, 1971.

Dubos, René. *So Human an Animal.* London: Rupert Hart-Davis, 1970.

Dumont, René, and Mazoyer, Marcel. *Socialisms and Development.* New York: Praeger, 1973.

Dwyer, Johanna T., and Mayer, Jean. "Beyond Economics and Nutrition: The Complex Basis of Food Policy." *Science* 188(9 May 1975):566–570.

Eckholm, Erik P. *Losing Ground: Environmental Stress and World Food Prospects.* New York: Norton, 1976.

Ehrlich, Paul R. *The Population Bomb.* New York: Ballantine Books, 1968.

Ehrlich, Paul R.; Ehrlich, Anne H.; and Holdren, John P. *Ecoscience: Population Resources, Environment.* San Francisco: Freeman, 1977.

Elliot, David, and Elliot, Ruth. *The Control of Technology.* New York: Springer-Verlag, 1976.

Falk, Richard A. *A Study of Future Worlds.* New York: Free Press, 1974.

Food and Agriculture Organization. *Provisional Indicative World Food Plans,* 2 vols. Rome: Food and Agriculture Organization, 1970.

Food and Agriculture Organization. *Annual Fertilizer Review, 1971.* Rome: Food and Agriculture Organization, 1972.

Food and Agriculture Organization. *Yearbook of Fishery Statistics, 1971,* Vol. 32. Rome: Food and Agriculture Organization, 1972.

Food and Agriculture Organization. *Trade Yearbook, 1974.* Rome: Food and Agriculture Organization, 1975.

Ford Foundation Annual Report. 1959–1975.

Frank, Jerome D. *Sanity and Survival.* New York: Random House, 1967.

Frankel, Francine R. *India's Green Revolution: Economic Gains and Political Costs.* Princeton, N.J.: Princeton University Press, 1971.

Frykenberg, Robert E., ed. *Land Control and Social Structure in India.* Madison: University of Wisconsin Press, 1969.

Fussell, George E. *Farming Technique from Prehistoric to Modern Times.* London: Pergamon Press, 1966.

Geertz, Clifford. "The Social–Cultural Context of Policy in Southeast Asia." In *Southeast Asia: Problems of United States Policy,* edited by William Henderson, pp. 45–70. Cambridge, Mass.: MIT Press, 1963.

Georgescu-Roegen, Nicholas. *The Entropy Law and the Economic Process.* Cambridge, Mass.: Harvard University Press, 1971.

Georgescu-Roegen, Nicholas. "Economics and Entropy." *The Ecologist* 2(July 1972):13–18.

Glacken, Clarence J. "Changing Ideas of the Habitable World." In *Man's Role in Changing the Face of the Earth*, edited by William L. Thomas, Jr., pp. 70–92. Chicago: University of Chicago Press, 1956.

Goulet, Denis. *The Uncertain Promise: Value Conflicts in Technology Transfer*. New York: IDOC/North America, 1977.

Grant, James P. "Energy Shock and the Development Prospect." In *The United States and the Developing World*, edited by James W. Howe, pp. 31–50. New York: Praeger, 1974.

Grant, James P. "Humanitarian Food Assistance in the New Era of Resource Scarcities." Washington, D.C.: Overseas Development Council, 1974. (Mimeographed.)

"Green Revolution (II): Problems of Adapting a Western Technology." *Science* 186 (December 1974):1186–1188.

Griffin, Keith. *The Green Revolution: An Economic Analysis*, Report No. 72.6. Geneva: United Nations Research Institute for Social Development, 1972.

Griffin, Keith. *The Political Economy of Agrarian Change*. Cambridge, Mass.: Harvard University Press, 1974.

Haas, Peter. "A Humanitarian Crisis." *Swiss Review of World Affairs* 24(May 1974):25–26.

Haggett, Peter. *Locational Analysis in Human Geography*. London: Edward Arnold, 1965.

Hannon, Bruce M. "An Energy Standard of Value." *The Annals* 410(November 1973):139–153.

Hansen, Roger D., ed. *The U.S. and World Development: Agenda for Action 1976*. New York: Praeger, 1976.

Havens, Thomas R. H. *Farm and Nation in Modern Japan*. Princeton, N.J.: Princeton University Press, 1974.

Heilbroner, Robert L. *An Inquiry into the Human Prospect*. New York: Norton, 1974.

Held, R. Burnell, and Clawson, Marion. *Soil Conservation in Perspective*. Baltimore: Johns Hopkins Press for Resources for the Future, 1965.

Hepting, George H. "Climate and Forest Diseases." In *Man's Impact on Terrestrial and Oceanic Ecosystems*, edited by William H. Mathews, Frederick E. Smith, and Edward D. Goldberg, pp. 203–226. Cambridge, Mass.: MIT Press, 1971.

Herendeen, Robert A. *An Energy Input–Output Matrix for the United States, 1963: User's Guide*. Urbana: University of Illinois, Center for Advanced Computation, Document No. 69, 4 March 1973.

Hesse, Herman. *Magister Ludi* [*The Bead Game*]. New York: Ungar, 1949.

Hewes, Laurence. *Rural Development: World Frontiers*. Ames: Iowa State University Press, 1974.

Heyerdahl, Thor. "How Vulnerable Is the Ocean?" In *Who Speaks for Earth?*, edited by Maurice F. Strong, pp. 45–63. New York: Norton, 1973.

Hightower, Jim, and DeMarco, Susan. *Hard Tomatoes, Hard Times: The Failure of the Land Grant College Complex*. Cambridge, Mass.: Schenkman, 1972.

The Holcomb Research Institute. *Environmental Modeling and Decision Making: The United States Experience*. New York: Praeger, 1976.

Horton, Robin, and Finnegan, Ruth, eds. *Modes of Thought: Essays on Thinking in Western and Non-Western Societies*. London: Faber and Faber, 1973.

Horvath, Janos. *Chinese Technology Transfer to the Third World: A Grants Economy Analysis*. New York: Praeger, 1976.

Horvath, Janos. "Food Production and Farm Size: A Reconsideration of Alternatives." Holcomb Research Institute Working Paper. Indianapolis, Ind.: Butler University, 1977.

Hoskins, William G. *The Making of the English Landscape.* London: Hodder and Stoughton, 1955.
Hoyle, Brian S., ed. *Spatial Aspects of Development.* New York: Wiley, 1974.
Hughes H. Stuart. *Consciousness and Society.* New York: Knopf, 1958.
Hughes, H. Stuart. *History as Art and as Science.* New York: Harper & Row, 1964.
Hutchinson, Joseph. "Crop Plant Evolution: A General Discussion." In *Essays on Crop Plant Evolution,* edited by Joseph Hutchinson, pp. 166–181. Cambridge: Cambridge University Press, 1965.
Hutchinson, Joseph, ed. *Population and Food Supply.* Cambridge: Cambridge University Press, 1969.
Illich, Ivan. *Tools for Conviviality.* New York: Harper & Row, 1973.
International Bank for Reconstruction and Development. *Rural Development: Sector Policy Paper.* Washington, D.C.: IBRD, 1975.
Jacks, Graham V., and Whyte, Robert O. *The Rape of the Earth: A World Survey of Soil Erosion.* London: Faber and Faber, 1939.
Jacobs, Alan H. "Maasai Pastorialism in Historical Perspective." In *Pastorialism in Tropical Africa,* edited by Théodore Monad, pp. 406–424. London: Oxford University Press, 1975.
Janzen, Daniel H. "Tropical Agroecosystems." *Science* 182(21 December 1973):1212–1219.
Jéquier, Nicolas, ed. *Appropriate Technology: Problems and Promises.* Paris: OECD Development Centre, 1976.
Johnson, Brian. *The Politics of Money.* New York: McGraw-Hill, 1970.
Johnston, Bruce F. "The Japanese 'Model' of Agricultural Development: Its Relevance to Developing Nations." In *Agriculture and Economic Growth: Japan's Experience,* edited by Kazushi Ohkawa, Bruce F. Johnston, and Hiromitsu Kaneda, pp. 58–102. Princeton, N.J.: Princeton University Press, 1969.
Jones, Eric L., ed. *Agriculture and Economic Growth in England, 1650–1815.* New York: Barnes and Noble, 1967.
Jones, Eric L., and Woolf, Stuart J., eds. *Agrarian Change and Economic Development.* London: Methuen, 1969.
Jordan, Robert S., and Weiss, Thomas G. *International Administration and Global Problems: An Analysis of the World Food Conference.* New York: Praeger, 1976.
Kahn, Akhter Hameed. *Reflections on the Comilla Rural Development Projects.* Overseas Liaison Committee Paper No. 3, Washington, D.C.: American Council on Education, March 1974.
Kautsky, John H. *Communism and the Politics of Development: Persistent Myths and Changing Behavior.* New York: Wiley, 1968.
Kellogg, Charles H. "Climate and Soil." In *Climate and Man: Yearbook of Agriculture,* pp. 265–291. Washington, D.C.: U.S. Department of Agriculture, 1941.
Koo, Anthony Y. C. *The Role of Land Reform in Economic Development: A Case Study of Taiwan.* New York: Praeger, 1968.
Kou, Leslie T. C. *The Technical Transformation of Agriculture in Communist China.* New York: Praeger, 1972.
Kuhn, Thomas S. *The Structure of Scientific Revolutions.* Chicago: University of Chicago Press, 1962.
Lamb, Hubert H. *The Changing Climate.* London: Methuen, 1972.
Landsberg, Hans H.; Fischman, Leonard L.; and Fisher, Joseph L. *Resources in America's Future,* Baltimore: Johns Hopkins Press, 1963.

Lappé, Frances Moore, and Collins, Joseph. *Food First: Beyond the Myth of Scarcity.* Boston: Houghton Mifflin, 1977.

Lee, John M. *African Armies and Civil Order.* New York: Praeger, 1969.

Lipton, Michael. *Why Poor People Stay Poor: Urban Bias in World Development.* Cambridge, Mass.: Harvard University Press, 1977.

Lovins, Amory. "Energy Strategy: The Road Not Taken?" *Foreign Affairs* 55(October 1976):66–96.

MacCulloch, John A. *The Childhood of Fiction.* New York: Dutton, 1905.

Makhijani, Arjun. *Energy and Agriculture in the Third World.* Cambridge, Mass.: Ballinger, 1975.

Man in the Living Environment. Report of the Workshop on Global Ecological Problems. Madison: University of Wisconsin Press, 1972.

Mannheim, Karl. *Ideology and Utopia.* New York: Harcourt, Brace, 1960.

Margalef, Ramón. *Perspectives in Ecological Theory.* Chicago: University of Chicago Press, 1968.

Marias, Julian. *Generations: A Historical Method.* Alabama: University of Alabama Press, 1970.

Martins, Herminio. "The Kuhnian 'Revolution' and Its Implications for Sociology." In *Imagination and Precision in the Social Sciences,* edited by Thomas J. Nossiter, Albert H. Hanson, and Stein Rokkan, pp. 13–58. London: Faber and Faber, 1972.

Marx, Leo. *The Machine in the Garden: Technology and the Pastoral Ideal in America.* New York: Oxford University Press, 1965.

Masefield, Geoffrey B. *A Short History of Agriculture in the British Colonies.* Oxford: Clarendon Press, 1950.

Masefield, Geoffrey B. *A History of Colonial Agricultural Service.* Oxford: Clarendon Press, 1972.

McHarg, Ian L. *Design with Nature.* Garden City, N.Y.: Doubleday/Natural History Press, 1969.

Mead, Margaret. *Culture and Commitment: A Study of the Generation Gap.* Garden City, N.Y.: Doubleday/Natural History Press, 1970.

Meadows, Donella H.; Meadows, Dennis L.; Randers, Jørgen; and Behrens, William W., III. *The Limits to Growth.* New York: Universe Books, 1972.

Meeker, Joseph W. "The Comedy of Survival." *The Ecologist* 3 (June 1973):210–215.

Mesarovic, Mihajlo, and Pestel, Eduard. *Mankind at the Turning Point.* New York: Dutton, 1974.

Milikan, Max, and Hapgood, David. *No Easy Harvest.* Boston: Little, Brown, 1967.

Montagne, Robert. "The Nation-State in Modern Africa and Asia." In *Comparative World Politics,* edited by Joel Larus, pp. 58–71. Belmont, Calif.: Wadsworth, 1964.

Moon, Parker T. *Imperialism and World Politics.* New York: Macmillan, 1926.

Moore, Barrington, Jr. *Social Origins of Dictatorship and Democracy: Lord and Peasant in the Making of the Modern World.* Boston: Beacon, 1966.

Morgan, Robert J. *Governing Soil Conservation: Thirty Years of the New Decentralization.* Baltimore: Johns Hopkins Press for Resources for the Future, 1975.

Mosher, Arthur T. *Creating a Progressive Rural Structure.* New York: Agricultural Development Council, 1969.

Mumford, Lewis. *The Myth of the Machine: Technics and Human Development.* New York: Harcourt, Brace and World, 1967.

Mumford, Lewis. "The Highway and the City," In *Environment and Society,* edited by Robert T. Roelofs, Joseph N. Crowley, and Donald L. Hardesty, pp. 208–219. Englewood Cliffs, N.J.: Prentice-Hall, 1974.

Myrdal, Gunnar. *Asian Drama: An Inquiry into the Poverty of Nations,* 4 vols. New York: Twentieth Century Fund, 1968.

National Academy of Sciences. *Productive Agriculture and a Quality Environment.* Washington, D.C., 1974.

National Academy of Sciences. *Plant Studies in the People's Republic of China.* Washington, D.C.: 1975.

Nettl, J. Peter. "The State as a Conceptual Variable." *World Politics* 20(July 1968):559–592.

Odum, Eugene P. *Fundamentals of Ecology,* 3rd ed. Philadelphia: Saunders, 1971.

Odum, Howard. *Environment, Power and Society.* New York: Wiley, 1971.

Ophuls, William. *Ecology and the Politics of Scarcity.* San Francisco: Freeman, 1977.

Oppenheimer, J. Robert. *The Flying Trapeze: Three Crises for Physicists.* London: Oxford University Press, 1964.

Organization for Economic Cooperation and Development. *Aid to Agriculture in Developing Countries.* Paris: OECD, 1968.

Overseas Development Institute. "Stimulating Local Development." Agriculture Administration Unit Occasional Paper. London: ODI, 1976.

Overseas Development Institute. "Extension, Planning, and the Poor." Agricultural Administration Unit Occasional Paper 2. London: ODI, 1977.

Owens, Edgar, and Shaw, Robert. *Development Reconsidered.* Lexington, Mass.: Lexington Books, 1972.

Paddock, William, and Paddock, Elizabeth. *We Don't Know How: An Independent Audit of What They Call Success in Foreign Assistance.* Ames: Iowa State University Press, 1973.

Paddock, William, and Paddock, Paul. *Famine 1975.* Boston: Little, Brown, 1967.

Palmer, Ingrid. *Food and the New Agricultural Technology,* Report No. 72.9. Geneva: United Nations Research Institute for Social Development, 1972.

Palmer, Ingrid. *Science and Agricultural Production,* Report No. 72.8. Geneva: United Nations Research Institute for Social Development, 1972.

Palmer, Ingrid. *The New Rice in the Philippines,* Report No. 75.2. Geneva: United Nations Research Institute for Social Development, 1975.

Papadakis, Juan. *Agricultural Potentialities of World Climates.* Buenos Aires: Libro de Edicion Argentina, 1970.

Pearse, Andrew. *An Overview Report,* Report No. 75/C.11. Geneva: United Nations Research Institute for Social Development, 1975.

Perkins, Dwight H. *Agricultural Development in China: 1368–1968.* Edinburgh: University Press, 1969.

Phillips, Claude S., Jr. "The Revival of Cultural Evolution in Social Science Theory." *Journal of Developing Areas* 5 (April 1971):337–370.

Pimentel, David; Dritschilo, William; Krummel, John; and Kutzman, John. "Energy and Land Constraints in Food Protein Production." *Science* 190(21 November 1975):754–761.

Polanyi, Karl. *The Great Transformation.* Boston: Beacon Press, 1957.

Reagan, Michael D. *The Managed Economy.* New York: Oxford University Press, 1963.

Report of the Panel on the World Food Supply. *The World Food Problem: A Report of the President's Science Advisory Committee.* Washington, D.C.: The White House, May 1967.

Report of the Study of Critical Environmental Problems (SCEP). *Man's Impact on the Global Environment.* Cambridge, Mass.: MIT Press, 1970.

Report of the Study of Man's Impact on Climate (SMIC). *Inadvertent Climate Modification.* Cambridge, Mass.: MIT Press, 1971.

The Rockefeller Foundation: President's Review and Annual Report. 1941–1975.

"The Role of the Social Sciences in Rural Development." New York: The Rockefeller Foundation, Working Papers, 1976.

Roos, Leslie L., Jr., and Roos, Noralon P. *Managers of Modernization.* Cambridge, Mass.: Harvard University Press, 1971.

Ross, Earle D. *Democracy's College.* Ames: Iowa State University Press, 1942. Reprint ed., New York: Arno Press and the New York Times, 1969.

Roszak, Theodore. *Where the Wasteland Ends.* Garden City, N.Y.: Doubleday, 1972.

Russett, Bruce R. *International Regions and the International System.* Chicago: Rand McNally, 1967.

Sahlins, Marshall D., and Service, Elman R., eds. *Evolution and Culture.* Ann Arbor: University of Michigan Press, 1960.

Sauer, Carl O. *Agricultural Origins and Dispersals.* New York: American Geographical Society, 1952.

Scheving, Lawrence E.; Halberg, Franz; and Pauly, John E., eds. *Chronobiology.* Tokyo: Igaku Shoin, 1974.

Schultz, Theodore W. *Economic Growth and Agriculture.* New York: McGraw-Hill, 1968.

Schumacher, Ernst F. *Small Is Beautiful.* New York: Harper Torchbooks, 1973.

Sewell, John W., ed. *The United States and World Development: Agenda 1977.* New York: Praeger, 1977.

Shen, Tsung-hau. *The Sino-American Joint Commission on Rural Reconstruction.* Ithaca, N.Y.: Cornell University Press, 1970.

Singer, J. David. "The Level-of-Analysis Problem in International Relations." In *The International System: Theoretical Essay,* edited by Klaus Knorr and Sidney Verba, pp. 77–92. Princeton, N.J.: Princeton University Press, 1961.

Soja, Edward W. *The Geography of Modernization in Kenya.* Syracuse, N.Y.: Syracuse University Press, 1968.

Southworth, Herman M., and Johnston, Bruce F., eds. *Agricultural Development and Economic Growth.* Ithaca, N.Y.: Cornell University Press, 1967.

Sprout, Harold, and Sprout, Margaret. *Towards a Politics of the Planet Earth.* New York: Van Nostrand Reinhold, 1967.

Sprout, Harold, and Sprout, Margaret. "An Ecological Paradigm for the Study of International Politics." Princeton, N.J.: Center of International Studies, Research Monograph No. 30, 1968.

Stakman, Elvin C., Bradfield, Richard; and Mangelsdorf, Paul. *Campaigns against Hunger.* Cambridge, Mass.: Harvard University Press, 1967.

Stavis, Benedict. *Making Green Revolution: The Politics of Agricultural Development in China,* Monograph No. 1. Ithaca, N.Y.: Cornell University Press, Rural Development Committee, 1974.

Steinhart, Carol, and Steinhart, John. *Energy: Sources, Use and Role in Human Affairs.* North Scituate, Mass.: Duxbury Press, 1974.

Steinhart, John, and Steinhart, Carol. "Energy Use in the United States Food System." *Science* 184(19 April 1974):307–316.

Stevens, Robert D. "Rural Development Programs for Adaptation from Comilla, Bangladesh." In *Agricultural Economics,* Report No. 215. East Lansing: Michigan State University, Department of Agricultural Economics, June 1972.

Streeter, Carroll. "Reaching the Developing World's Small Farmers." New York: Rockefeller Foundation, Working Papers, 1972.

Tawney, Richard H. *Religion and the Rise of Capitalism.* London: Murray, 1926.

Thomas, Franklin. *The Environmental Basis of Society.* New York: Century, 1925.

Thomas, William L., Jr., ed. *Man's Role in Changing the Face of the Earth,* 2 vols. Chicago: University of Chicago Press, 1956.

Thorson, Thomas L. *Biopolitics.* New York: Holt, Rinehart & Winston, 1970.

Tinbergen, Jan (coordinator). *RIO: Reshaping the International Order.* New York: Dutton, 1976.

Tinker, Jon. "The End of the English Landscape." *New Scientist* 64(5 December 1974):722–727.

Tocqueville, Alexis de. *Democracy in America,* 2 vols. New York: Vintage Books, 1954.

Tocqueville, Alexis de. *Journey to America.* New Haven, Conn.: Yale University Press, 1959.

Toma, Peter A. *The Politics of Food for Peace.* Tucson: University of Arizona Press, 1967.

Toynbee, Arnold J. *A Study of History.* New York: Oxford University Press, 1948–1961.

Toynbee, Arnold J., in collaboration with Jane Caplan. *A Study of History,* latest revised and abridged version. London: Oxford University Press in association with Thames and Hudson, 1972.

Tuan, Yi-Fu. *The Hydrologic Cycle and the Wisdom of God.* Toronto: University of Toronto Press, 1968.

Tuma, Elias H. *Twenty-six Centuries of Agrarian Reform.* Berkeley: University of California Press, 1965.

Turner, Louis. "Multinationals, the United Nations, and Development." *Columbia Journal of World Business* 7(September/October 1972):13–22.

Turner, Louis. *Multinational Companies and the Third World.* New York: Hill and Wang, 1973.

Tuveson, Ernest L. *Millenium and Utopia: A Study of Background of the Idea of Progress.* Berkeley: University of California Press, 1949.

United Nations. *Natural Resources of Developing Countries: Investigation, Development, and Rational Utilization.* New York: United Nations Dept. of Economic and Social Affairs, 1970.

United Nations. *World Plan of Action for the Application of Science and Technology to Development.* New York: United Nations Dept. of Economic and Social Affairs, 1971.

United Nations Educational Scientific, and Cultural Organization. *Status and Trends of Research in Hydrology, 1965–1974.* Paris: UNESCO, 1972.

United Nations Research Institute for Social Development. *The Social and Economic Implications of Large-Scale Introduction of New Varieties of Food Grain: Summary of a Global Research Project,* Report No. 74.1. Geneva: UNRISD, 1974.

Vallianatos, Evan. *Fear in the Countryside: The Control of Agricultural Resources in the Poor Countries by Nonpeasant Elites.* Cambridge, Mass.: Ballinger, 1976.

Vann, Anthony, and Rogers, Paul, eds. *Human Ecology and World Development.* New York and London: Plenum Press, 1974.

Wallerstein, Immanuel. *The Modern World System: Capitalist Agriculture and the Origins of the European World-Economy in the Sixteenth Century.* New York: Academic Press, 1974.

Wan, Hsuing. "Agricultural Research Organizations in Taiwan." In *National Agricultural Research Systems in Asia,* edited by Albert H. Moseman, pp. 89–98. New York: Agricultural Development Council, 1971.

Ward, Benjamin. *What's Wrong with Economics.* New York: Macmillan, 1972.

Ward, Ritchie R. *The Living Clocks.* New York: Mentor Books, 1971.

Warriner, Doreen. *Land Reform in Principle and Practice.* Oxford: Clarendon Press, 1969.

Watson, James D. *The Double Helix.* New York: Atheneum, 1968.

Weber, Max. *The Protestant Ethic and the Spirit of Capitalism.* New York: Scribner, 1958.

Wharton, Clifton R., Jr., ed. *Subsistence Agriculture and Economic Development.* Chicago: Aldine, 1969.

Whitcombe, Elizabeth. *Agrarian Conditions in Northern India: The United Provinces under British Rule 1860–1900.* Berkeley: University of California Press, 1972.

Whitcombe, Elizabeth. "The New Agricultural Strategy in Uttar Pradesh, India, 1968–70: Technical Problems." In *Technical Change in Asian Agriculture,* edited by Richard T. Shand, pp. 183–201. Canberra: Australian National University Press, 1973.

White, John. "What Is Development? And for Whom:" Given at the U.K. Quaker Conference on "Motive Force in Development," April 1972. (Mimeographed.)

White, Lynn, Jr. "The Historical Roots of our Ecological Crisis." In *The Subversive Science,* edited by Paul Shepard and Daniel McKinley, pp. 341–351. Boston: Houghton Mifflin, 1969.

Wightman, David R. "Food Aid and Economic Development." *International Conciliation,* No. 567, March 1968.

Wilkes, Garrison, and Wilkes, Susan. "The Green Revolution." *Environment* 14(October 1972):32–39.

Wittfogel, Karl A. *Oriental Despotism.* New Haven, Conn.: Yale University Press, 1957.

Wittfogel, Karl A. *Agriculture: A Key to the Understanding of Chinese Society, Past and Present.* Canberra: Australian National University Press, 1970.

Woodham-Smith, Cecil. *The Great Hunger.* London: Hamish Hamilton, 1962.

Zorza, Victor. "Will Mao's Cupboard Be Bare?" *Guardian Weekly* 23(November 1975):11.

Index